Elizabeth Kasmer has been an exchange student, a donor attendant, a receptionist, a carpark attendant, security shredder and a primary school teacher. She has always loved books, frequenting her local library as a child. She was enthralled by CS Lewis, embarrassed by Judy Blume, delighted in Roald Dahl, discovered the secret to 'Life, the Universe and Everything' and giggled along with Sue Townsend's dorky and adorable Adrian Mole aged 13 and ¾. She currently lives on the Sunshine Coast with her husband, three mini ninjas (cleverly disguised as boys), a dog and a snake. *Becoming Aurora* is her first novel.

elizabethkasmer.com

BECOMING AURORA

BECOMING AURORA

ELIZABETH KASMER

UQP

First published 2016 by University of Queensland Press
PO Box 6042, St Lucia, Queensland 4067 Australia
Reprinted 2018, 2020

www.uqp.com.au
reception@uqp.com.au

Cover design by Jo Hunt
Cover photographs: girl by HyperionPixels/iStock; railway by bravo1954/iStock; clouds by elenavolkova/iStock
Author photograph by Larissa Salton
Typeset in 12/16 pt Adobe Garamond by Post Pre-press Group, Brisbane
Printed in Australia by McPherson's Printing Group

Cataloguing-in-Publication Data
National Library of Australia
http://catalogue.nla.gov.au

Kasmer, Elizabeth, author.
Becoming Aurora / Elizabeth Kasmer.

ISBN 978 0 7022 5420 8 (pbk)
ISBN 978 0 7022 5587 8 (pdf)
ISBN 978 0 7022 5588 5 (epub)
ISBN 978 0 7022 5589 2 (kindle)

University of Queensland Press uses papers that are natural, renewable and recyclable products made from wood grown in well-managed forests and other controlled sources. The logging and manufacturing processes conform to the environmental regulations of the country of origin.

For Cliff,
cynic, pirate, librarian,

and for Jenny,
who taught me the true meaning of the word vocation.

Tonight we are wolves.

Our pack moves as one, past empty shop fronts and faded billboards. On the corner we gather in the shadows of the Royal George Hotel. Last drinks were served hours ago, but the stink of stale beer lingers. The signal box on the corner clicks to itself and the traffic lights cycle through from green to red and back. We sprint across an empty road in search of our prey. Rail tracks from the abandoned sugar mill gleam yellow in the streetlights and curve through the roads of our town, leading to nowhere.

We crowd the doorway of the Curry House, bristling with excitement. One of us steps forward and, taking aim with a cricket bat, smashes the bottom half of the plate glass door. Heavy boots kick off the sharp edges and we pile in through the opening. Inside the restaurant our game is played by shadows. Chairs and tables are thrown, tacky ornaments souvenired, everything is fair game. In the kitchen someone upends a huge sack of rice. Fragrant jars of red and yellow spice are hurled into the air as we dance like wild things at a celebration. Our message sprayed in red paint across the white walls of the restaurant.

GO BACK TO WHERE YOU CAME FROM

ONE

On my seventh birthday my father took me to visit a goddess and, when I think back, three perfect and distinct memories come to mind.

The first is of the train rocking and jolting past overgrown backyards and pineapple plantations. It slows down and we snake our way past what looks like a giant stone gorilla hunkered on the ground. The gorilla's sullen face is frozen: forever staring off into the distance. Quick breaths fog the window. *It's only Mt Tibrogargan,* my father explains.

The second is being caught in a spell as my father guides me through the silence of the gallery to meet her.

Finally, on the journey home, hearing stories of how she bewitched artists and poets. How each morning the birds lift their heads from under their wings to herald her coming, calling out her name in a long, unbroken chain.

Aurora.

*

It's the first week of the summer holidays and already I can't wait for them to be over. Mid-afternoon it's still stinking hot, but a breeze makes the walk to the park bearable. Judging by the number of people spread out under the trees, everyone must be here. I make my way up the grassy slope and rake my fingers through my hair.

Cam's propped up against the trunk of our tree, the one we climbed in Year 4 on a dare so that we could look out through the branches to the world below. It's the tree with our names carved into it, in a secret place inside its ancient folds. He spots me first and calls out for me to join him. Heads turn in my direction and the rest of them start to clap and cheer.

He's wearing a pair of cut-off jeans, and a t-shirt that reads: *Don't Tase Me, Bro.* Shorts are really a bad choice for Cam. His legs are long, hairy and blindingly white; but no one has the heart to tell him that he looks like an out-of-control spider.

'So, how'd the meeting with the director go? Is community service shaping up like a hundred-hour episode of *Survivor*, or more,' he tilts his head, searching for the right words, 'like a piece of cake?'

'Sounds pretty easy – gardening, cleaning and generally helping out wherever it's needed. It's doable.' I shrug.

He lies back on the grass and folds his hands behind his head. 'I know it sucks to be you right now, but it was *so* worth it.' Cam picks a long piece of grass, chews the end, smiling. 'You've got guts, Rory. I'm impressed. We all are.'

I smile and stretch out in the dappled sunlight. For the first time in a long time I feel at peace. 'Hey, what are you doing tomorrow? Maybe I can sneak you in so you can help me with the weeding.' Cam screws up his face and I give him a playful shove with my foot. We muck around for a while before I realise the time. 'I've gotta pick up Eddie by 5.30.'

I make my way through the group, smiling as my friends slap me on the back, and my world seems to grow that much bigger.

The day-care woman answers the door and immediately I know I'm late. She doesn't look me in the eye. Eddie barrels into my legs and holds on tight.

'Ory, Ory, we go home now?' Eddie's baby cheeks are bright red. His dark hair is plastered to his forehead.

'He's done nothing but run around out the back chasing the chickens. I don't know how he can in this heat.' She hands me his backpack. 'Tell your mum he's had a good nap this afternoon, about two hours, and he ate all his lunch.' She squats down so she's at eye level. 'You've been so good today, Eddie. I'll see you in a few days, okay?' She plants a kiss on his sweaty cheek and stands to look at me. Her face hardens. 'Your mum's working an early shift on Tuesday?'

I nod.

'Okay, Eddie, I'll be seeing you on Tuesday, then.' She waves him goodbye and the screen door snaps shut. Eddie keeps waving as I swing his backpack onto my shoulders. His tiny hand reaches out for mine, and I feel an invisible tug. We walk side by side, keeping to the shade along the edge of the footpath. I give his little hand a squeeze. It's sticky, but I don't want to let it go.

'Look.' He's pointing his stubby finger at something across the road. I look really hard and then I see it. A tiny willy-wagtail darting in and out of the trees. I watch as it waves its black fan-shaped tail in our direction. I used to love those little birds. Not anymore.

'Come on.' I tug on his arm, but he doesn't move. He's fascinated by the bird's jerky, unpredictable movements. I grab onto his shirt and half push, half steer him down the pavement. Eddie's too wrapped up in the thrill of the walk to notice that the bird's following us. We round the corner. I can still see it out of the corner of my eye. It looks as if it's leaning forward, chirping right at me, and suddenly I'm drifting, fading away.

'Get lost!' I yell.

TWO

I'm in the kitchen when the oven timer goes off. Somehow I've managed to burn the chips again. I pick out the black ones, toss them into the bin and head to the bathroom.

'Come on, time to get out!' I yell down the hall.

Eddie's singing in the shower when I lean in and turn off the taps.

'No!' he squeals.

But despite his protests I manage to drop the towel over his head and rub his wet hair. 'It's your favourite: hot chips and sausage rolls.'

He pulls the towel off his head, sizing up my offer. 'Sauce too?'

Before I can answer he's running down the hallway, naked and dripping wet. A few minutes later he reappears, plonking himself on the edge of the couch, his pyjama shirt on, even if it is inside out and back to front. He upends the sauce bottle onto his plate and shakes hard.

So much comes out that it completely covers his chips. I roll my eyes and we laugh.

After dinner we lie on his small bed reading *Green Eggs and Ham* until the wriggling stops and is replaced by slow, steady breaths. The ceiling fan hums intermittently to itself and, despite the heat, his little body is pushed up hard against mine. In this light, Eddie reminds me of the angels in Dad's art history books. Big, heavy books filled with fat cherubs floating about on wings and endless paintings titled *Madonna and Child.*

The front door opens and there's the sound of keys being dropped on the bench. Mum appears in the bedroom doorway, her nurse's photo ID still clipped to her shirt. She comes over and kisses Eddie on the head, but it's obvious she's pissed off. My legs are tangled in the sheets and I slowly pull myself free and head outside. I'm not in the mood for a fight.

It's quiet on the back steps. A full moon's rising behind the mango tree. The silhouette of a possum wanders across a lower branch and stops at the bird feeder. The moon casts a yellow glow onto the garage. Its gutters are full of leaf litter and the paint is peeling from under the eaves. I'd love to sand it back, give it a fresh coat of paint, but Mum flat out refuses to discuss it.

Music drifts from the kitchen window. It's their song, the one she plays when she's missing Dad. The music in the house is turned up and I mouth the words.

*

The phone's ringing in the kitchen. I open bleary eyes, trying to focus on the face of my bedside clock. Who in their right mind would be calling this early on a Sunday? I roll over in bed and rub my sore, stiff shoulders, a consequence of yesterday, my first day weeding at St Mary's. A voice murmurs from down the hall and then Mum's at my door. 'Rory, the phone is for you. It's the director of nursing.' Her voice has a tight edge to it, as if she's preparing herself for more bad news.

I rub the sleep from my eyes and shuffle past. Mum presses her fingers into the sides of her forehead as she follows me into the kitchen. I take a deep breath and place the phone to my ear.

A familiar voice snaps down the line. For a moment I stand a little straighter in case she somehow knows I'm still in my pyjamas, which is stupid, so I consciously relax. Her voice is so loud I have to hold the phone away from my ear: something about the work I did yesterday not being up to scratch. She doesn't wait for an answer; what I have to say doesn't matter to her in the slightest. Finally she pauses for breath, no doubt hoping this will help her message sink in. But all I'm thinking is that this woman has officially cracked my top-five-annoying-people-of-all-time list.

'Mr Sanford has asked for someone to read to him. He's in the high care ward. It's Sanford, S-a-n-f-o-r-d.' She talks slowly, as if I'm some sort of idiot. 'From now on, regardless of whether you have finished your set tasks on time, you will be required to spend at least half an hour reading to Mr Sanford. If I'm not satisfied that you are doing your best, I will be forced to contact the magistrate directly.'

The line goes dead and I'm cast adrift in a sea of endless beeps.

'Anything I need to know?' Mum asks. She's standing with her back to me, stirring sugar into two mugs on the bench. But from the tone of her voice, she really doesn't want to know. I grab one of the mugs, take it back to my room and close the door behind me. My phone buzzes. It's a text from Cam inviting me for a swim at his house 'after' – which I assume means after serving time at St Mary's. Cam has a habit of being vague. I text back with instructions to meet me at the back entrance of St Mary's if he wants to help with the weeding. Of course, there's no answer.

I've almost survived my second shift when I arrive at an intimidating set of double doors. Printed across the top is a sign telling me I've arrived at the High Care Ward. A nurse sits behind a large workstation. The badge on her shirt says her name is Fiona. She smiles at me over the top of her glasses.

'Ah, the director of nursing sent me. She said I'm to read to Mr Sanford,' I say.

Fiona sits her glasses on top of her head. 'Ah yes, the infamous Jack Sanford. You must be Rory. The director just rang to let me know you were on your way. Efficient, isn't she?'

I catch the sarcasm in her voice and we smile at each other as if we're sharing a private joke.

She leans over the counter and points with her pen. 'Jack's room is the second door on the right.'

I stride down the hall and push the heavy door open with both hands. It's dark inside and the air smells of decay, like the apple core I forgot about in the bottom of my bag. An elderly man is lying motionless on the bed. His eyes are closed and his mouth hangs slightly open. My eyes adjust to the darkness and there is only the sound of steady breathing. This room is nothing like the rooms of the tongue-clicking oldies down the hall. Theirs are bursting with family portraits, plants, paintings, anything to remind them of the life they once had. The only thing on the walls of this room is a sad Jesus hanging from a plastic cross.

I turn to leave and open the door as quietly as I can.

'What do you want?' a gravelly voice rasps behind me.

I hold the door for a second too long, trying to find the right place to start.

'I said: what-do-you-*want*?'

'My name is Rory. The director of nursing sent me. She said I have to read to you.'

The old man looks me up and down, taking in my black jeans, black sandshoes and a dirt-smeared, sweat-stained t-shirt. 'You don't look like someone who can read.' And even in the darkness I can see him sneer.

By the time I was 12, I'd read every book in the children's section of my local library. The librarians used to joke about it every time I came in with an armful of returns. But that was years ago. That was before.

'Look, girlie, I don't want the likes of you wasting what precious little time I have left. Why don't you go back to your do-gooders group and leave me alone!'

I roll my eyes and snort. 'Like lying in here in the dark is *such* a good use of your time.' I pull the blinds open. The sudden light leaves both of us blinking. He shades his eyes and curses under his breath. He's older than I thought at first. Stick-thin with a shock of white hair that stands up at odd angles. His faded pyjamas are spotted with brown food stains and big bags of flesh hang under his eyes. He reaches across to put on a pair of heavy, framed glasses.

'Just who do you think you are?' His eyes flash with anger, but it passes quickly. 'Just go.'

As I walk by the nurses' station Fiona closes a manila folder and places it carefully on the top of her in-tray.

'I tried,' I say, shrugging.

'It's Aurora Morris, isn't it?'

I nod, suddenly suspicious.

'Your father was my daughter's art teacher at school. I don't know how he did it, but he managed to keep my Lara on the straight and narrow. He was a good man.' Her voice softens. 'I was really sorry to hear of his passing.'

A nod is all I can manage. It's been four years since Dad died and I still can't talk to strangers about him. It doesn't surprise me that he helped Lara. He was always helping someone. Dozens of students came to his funeral. I sat in the front row and watched them. The younger ones in uniform, the older ones dressed in suits and ties moving forward in a slow procession, each one placing an origami heart on his coffin. I was told later that when he handed out detentions the offending students were ordered to fold a bucketload of red origami hearts. Dad always said paper folding was a lesson in meditation and far better than dishing out corporal punishment.

That was just like Dad.

I check my phone as I step from the foyer of St Mary's and the heat hits me square in the face. The wind's gusting from every direction. I blink as clouds of leaves and dirt sting my eyes. I walk across town to Cam's house, pausing every now and then to shield my face from the barrage of dust and dirt that's being picked up and flung across the street by an invisible hand.

By the time I arrive, beads of sweat have formed on my top lip and my shirt's sticking to my back. Ellen, Cam's grandma, answers the door, her broad smile a welcome sight. She's taken care of Cam since he was just a baby. Her large frame fills the door and she's wearing one of those loose floral dresses that women her age seem to like. Steel-grey hair is combed back off her head, revealing a sharp widow's peak.

'Rory, my beautiful girl.' She opens her arms and wraps me in a bone-crushing hug. I breathe in deeply, enjoying her cigarette and breath-mint scent. She holds me at arm's length and looks at me carefully. 'How are ya, love?' There's such a deep seriousness to the tone of her voice that, for a moment, I feel like she can see right inside me. Images flash across my mind of all the things I have folded neatly away: the bird, the old man, the director, my mum. Then she blinks. And it's just us again.

'Ellen, I'm fine, everything's fine.' I hug her again, lingering in the warmth and familiarity of her embrace. 'But thanks for asking.'

'That's too many "fine"s in one sentence.' She smiles back and I can see the dark shadows in her mouth, the gaps where her teeth are missing.

Outside, there's a loud splash and I follow the sounds through the house to the backyard. The above-ground pool is heaving as water slaps at its sides and gushes over the edge. I don't even stop to say hello as I wriggle out of my dirty, sticky clothes down to my knickers and bra and jump in. I sink to the bottom and glide through the cool quiet. A pair of legs comes towards me. Cam ducks under the water and pulls faces at me before turning on himself in an exaggerated, slow-motion battle. I giggle until my lungs burn, push off the bottom and burst through the surface, washed clean.

Cam doesn't even stop to look up from his mock battle with himself.

'Rory, Hulk versus Spiderman: who would win?' I know this isn't a serious question, but I love him all the same.

'You're such an idiot.' I roll my eyes and splash him.

'Yeah, you're right, it's not really much of a competition,' he says. 'Everyone knows the Hulk'd kick Spiderman's butt.'

Ellen appears on the back deck with a tray full of biscuits and two cans of soft drink. She sits on a faded plastic chair in the shade to smoke a cigarette. Cam leaps over the side of the pool and grabs most of the biscuits before she pushes his dripping body away and hands me the plate with the last three on it – and they're soggy.

I hadn't realised how hungry I am. It's been a long day.

She coughs and stubs her cigarette out in the plastic ashtray overflowing with grey ash and squashed butts. Her chair scrapes the ground as it's pushed back. She stands, her eyes glued to the inside of my ankle.

'Rory, what on earth were you thinking? I half expected this one might be stupid enough, but, honest to God, Rory, I really thought—' Her voice trails off, but her eyes stay focused on the small tattoo on my ankle. Cam gives me a sideways smile. The tattoo's still ringed by angry redness and I resist the urge to scratch at the scab with my other foot. I study the image. The stars of the Southern Cross are inked just above the arch of my foot. I remember the sting of the tattooist's needle and how I secretly enjoyed the feeling that, in some way now, I belonged to the gang, that I shared their mark.

Ellen's eyes seem drawn with defeat and sadness, but I can't bear the thought of disappointing her; nothing I can say will help her understand.

Cam unconsciously runs his hand over his own matching tattoo. His is much bigger than mine and sits on the left side of his chest, so it could be close to his heart. 'Gran, Rory earned that tat, just like I did.'

'Don't you dare drag this girl into your mess.' The flesh that hangs from her arm shakes with rage as she thrusts an accusing finger at him. Sweat beads on her top lip and she has to stop to catch her breath.

She leans against the chair for support. She's wheezing: sucking in great lungfuls of air before letting loose a volley of hacking coughs and I shrink a little further into the curve of my plastic chair.

THREE

Nurse Fiona looks up from behind the desk and checks her watch. 'Rory, you're right on time. The director of nursing *will* be happy.'

'Oh, goody.' I smirk. 'I've decided to dedicate my life to trying to make the director happy.'

Fiona winks. 'Probably best you keep that gem to yourself.'

Rule number one for getting through community service: always be nice to the woman behind the desk. Should be easy; I'm starting to like Fiona more and more.

'I don't know what she told you about Jack, but he's pretty sick. Not that he'd admit it: there's still a bit of fight left in the old fella.' Fiona clicks the end of her pen and shoves it in her pocket. 'He told me the other day he used to be part of a boxing troupe when he was a young bloke. Like a travelling show that toured around outback towns. Just think of the stories he could tell, eh?'

A loud crash echoes down the hall and she sighs before she hurries off to investigate.

I open the door to Jack's room and almost collide with a middle-aged woman in a green uniform. She's holding a cleaning cloth in one hand and a spray bottle filled with pink liquid in the other.

She checks her watch and clicks her tongue. 'Nearly done, love; just have to give the toilet a quick clean.' She motions for me to come in, and disappears through the doorway to Jack's small bathroom. My nose wrinkles at the strong smell of eucalyptus disinfectant that cuts the air. Jack rolls over in his bed and squints in my direction.

'Bloody hell, what does a man have to do to get a little privacy around here?'

I clear my throat. 'Mr Sanford, the director of nursing has asked me to read to you again.' I point to the backpack that hangs from my shoulder. He puts on his glasses and studies my face like he's trying to remember something that happened a long time ago.

'You,' he scoffs. 'Are you thick or something? I thought I told you to leave me alone.'

'All done,' a cheery voice calls from the doorway. The woman in the green uniform stacks her buckets and leaves.

'Here, why don't you take this one with you?' Jack points a bony finger in my direction and a sudden flash of anger washes over me.

'Look, I don't *want* to be here. This is not exactly how I'd choose to spend my holidays.' My life may be in the toilet, but I'll be damned if I'm going to let this old bastard get the better of me.

'Well, that makes two of us. Do you think I'd choose to be rotting away in this godforsaken room?' He folds his arms across his chest and glares at me.

I move my backpack from one shoulder to another. What I really want to do is leave, to be somewhere, anywhere, other than here. But I know the director will take great pleasure in reporting my lack of co-operation to the magistrate.

'So, are you going to tell me why you're really here?' Jack says.

I cross my arms and glare out the window, pressing my lips together. *You can get stuffed.*

The old man's face wrinkles with frustration and he wipes little strings of saliva from the side of his mouth. 'Hell's bells, I knew as soon as they stuck me in here it was all wrong. It's got so a man can't get any blessed rest. There's always someone wanting something. They're either poking at me or sticking me with those bloody needles. And, to top it all off,' he throws his hands up towards the ceiling, 'they send in a slacker like you!' He glares at me before dropping his arms to the bed and slumping back on the pillows, exhausted.

I shuffle and glance at the open doorway. At this

point, facing the wrath of the director can't be any worse than staying here.

'Oh, for God's sake, don't just stand there like a ninny.' He points to the vinyl chair in the corner. 'Let's just get this over with.'

I drag my eyes from the doorway and lumber around the bulky hospital bed to the visitor's chair. I reach into my backpack and pull out a tattered copy of *Wuthering Heights* that I've borrowed from Mum's bookshelf. I almost explain that after studying it in English last year, it's one of my favourite books, but his eyes are already half-closed. He unconsciously flexes his right hand, rubbing it. I look at his heavy-set jaw and imagine him fighting, taking a hit, welcoming it. Under the sickness I can see the mongrel in him.

Annoying Christmas carols play on the speaker system of the local shopping centre, but at least it's air-conditioned. The place has been decorated with sparkly tinsel for months now, but, judging by all the screaming kids, no one's having much fun. I'm sending a text when suddenly my head's jerked back. Some idiot's pulled hard on my ponytail. I spin around, but it's only Cam and he's just shoved an entire muesli bar into his mouth. He's chewing with his mouth open, pushing the food around with his tongue so that bits of oat fall onto the ground.

'Jeez, you're gross, *and* you're a pain in the bum. I think it's time for Ellen to have a heart-to-heart about your lack of table manners,' I say.

'Nah, Gran doesn't mind. She says I'm perfect and she loves me just the way I am.'

'Yeah, right,' I scoff, 'that *so* sounds like something she'd say.' I pull out my hair band and run my fingers through my long, dark hair. 'Oh, and the next time you think about pulling on my ponytail, just remember, I'll return the favour by pulling even harder on your balls, okay?'

We make our way through the crush of mums and kids with bulging shopping bags trying to beat the Christmas rush, when someone from our group calls out from the crowd.

'Hey, Cam, you heading to the park?' He turns in the direction of the voice and continues to walk backwards, bumping into everyone. He's nodding and giving the thumbs-up sign and the boy raises his chin and calls back, 'See you there, man.'

Away from the chaos of the shopping centre, crows call mournfully from the trees nearby, waiting for the road to empty so they can feast on the food scraps that litter the grounds. We teeter on the kerb and are buffeted by the whoosh of trucks and cars. There's a quick break in the traffic and we race across the first two lanes, stopping on the wide median strip as traffic rushes past from the opposite direction. The local council has recently been busy planting the median strip with small shrubs and fresh pine bark in a vain attempt to make our town less ugly. The tip of my shoe digs a hole in the bark as I watch and wait for the lights further up the road to turn red.

'Who'd have thought this sad excuse for a road used to be the main highway?' I say, almost to myself.

'Yeah, right. Gran told me about it when she took me to the Big Pineapple a few weeks ago. She heard talk on the radio that they might be shutting it down soon.'

'What? You've lost it. What's the Big Pineapple got to do with the highway?'

'Well, I asked Gran why they didn't think to build it on the main highway and then more people would come and it wouldn't be shutting down, and she said that it *was* built on the main highway. Till they put in the bypass.'

'Hang on, I thought you hated the Big Pineapple. Didn't you used to call it a tourist trap?'

'Well, yeah, but that was before all this talk of it closing down.'

'Cam Baker, who knew you were such a softie?' He can probably hear the mock admiration.

'What?' he says all defensive. 'Gran loves a fruit parfait – you know, like the Pineapple Boat?'

'Technically, the Pineapple Boat is not a parfait.'

'No, it's a culinary treat – that's what it is! Whoever thought of cutting a pineapple in half and filling it with fruit, ice-cream and cream is a certified genius.' He stares off into the middle distance, and his mouth hangs open. 'I could go one of them right now.'

The lights further up turn red and I give him a quick whack in the stomach to let him know it's clear for us to cross. He snaps back to attention and we jog across, our bags clunking against our backs. We skirt the edge of the road around a bend that passes directly under the railway bridge. The tail of a cattle train rumbles overhead. Whenever they pass by I try not to think too hard about what's waiting for the poor old cows at the end of the line.

I peer over the railing into Petrie Creek, the reason why a railway bridge was needed. Anytime we get decent rains in this town, the creek's where it floods first. Rubbish lines the banks and a shopping trolley is half buried in the reeds and mud. It's in desperate need of a good flushing out.

We cut through the open paddock up the hill towards the basketball courts. A group of boys is huddled on their usual concrete bench. One of them has brought a laptop and they're all leaning in, hooting with laughter.

'Hey.' The boy with the laptop motions us over. 'You've really got to see this. It is hil-ar-ious.'

We toss our bags at the base of the tree that towers above us and crowd round as he plays the video again from the beginning. Onscreen the picture bumps and shakes and then the image steadies. There's a kid about our age and he's sitting in a train carriage. He looks pretty normal in a t-shirt, shorts and a pair of those basketball shoes that go up high above the ankle. His long arms and legs are wrapped around the seat at a strange angle and he's twisting, trying to get a better view of someone behind him offscreen. By the way he's moving I'd say he's really drunk. He cups his hand around his ear and stretches precariously over the back of his chair.

'What's that? I can't understand you. You need to speak English.' He turns to the person holding the camera, laughing hysterically and slapping his leg. The camera pans to the back of the carriage and there's a man in a railway security uniform. These guys can often be seen hanging round Nambour station waiting for the next train. But this guy's different. His skin's the darkest shade of black I've ever seen. The train's bright fluorescent light reflects off his shiny face.

'Man,' says Cam, 'that's one black dude.'

The boy in the video continues. 'Why should I listen to you? You're not even Strayan – if you come, and esspect to live in *my* country.'

'I can understand you. I speak English just fine,' the guard replies in a thick African accent.

At this point everyone falls about laughing as they take turns trying to mimic the guard.

Onscreen laughter echoes in the background, and the picture wobbles uncontrollably. I get a weird sense these boys are laughing along with us. The camera now pans down the carriage to the shocked faces of the other passengers. Sitting nearby is a blonde girl. Her hair has been pinned up with a few loose strands framing her face. Her mouth hangs open in the shape of an 'o', like she can't quite believe what's going on.

The strange thing about this scene isn't the boy carrying on – it's the shocked looks on the faces of the passengers. They all just sit there like zombies.

I tap Cam on the shoulder and indicate that I'm off home. If this is how the rest of the afternoon's going to be played out I'm just not in the mood for it today.

'I've got to be home before 4 to look after Eddie,' I lie.

'But, Rors,' his voice is thick with disappointment, 'we just got here.'

I don't add anything else. I've learnt over the years that if you're going to lie, keep it simple. Anyone else and he'd insist on them staying. But he knows not to give me grief. He knows he owes me big time.

Cam follows and tries to link arms with me, but I shake him off. 'Hey, Rory, what's up? That video was supposed to be funny – you know, haha?' He's trying to sound like he's joking but I can tell he's not. 'Why do you have to spoil it by being such a girl?'

Anger bubbles up from deep inside. I stop on the edge of the footpath and turn to face him. 'I *am* a girl, Einstein. You guys just don't get it, do you? No one else had to front up to that stupid magistrate—' I struggle to find the right word, '—bloody judging me. Me, like I was some sort of lowlife scum, like that idiot on the video.' I'm all breathless and my heart's beating like crazy. Saying the words out loud has opened a floodgate of anger and resentment. 'Not one of you came round or rang to ask how it was going – how *I* was going.'

Cam looks surprised as he places a hand gently on my shoulder. 'You never said anything before. Besides, you know we weren't supposed to contact you or go anywhere near you during all that court appearance stuff – you know, for our protection.'

'And how do you think that *made me feel*?' I blink hard at the red hot stinging in my eyes.

'But—' He looks truly lost. 'You agreed to it; you said it was okay.'

Cam's mouth hangs open like the mouths of those zombies on the train. I cross my arms over my chest and stomp off in the opposite direction.

Right now, I need to be alone.

FOUR

It's Friday afternoon and Eddie and I are on the couch watching television when Mum rustles through the door, lugging plastic bags filled with groceries.

'Rory, don't forget the Christmas markets are on tonight.'

I stare at her blankly.

She puts the heavy bags on the table and starts unpacking them. 'You promised you'd come.' She looks at me with that I-can't-believe-you-forgot expression.

I know this is just a ploy to stop me from meeting up with Cam, but now Eddie's grabbed my arm and he's jumping up and down with enthusiasm. I sigh. If I back out now, he'll be as grumpy as only a four-year-old can. Besides, we'll be home by 7.30 at the latest, allowing plenty of time for other, more appropriate, Friday evening activities. 'Okay, little one,' I say, 'but I get to pick what we have for dinner.'

'Yay!' His hands shoot up in the air and he does a little victory dance. 'Big Ger-man sausages?'

I tickle him under the arms and he squeals with pleasure. 'That's *just* what I was about to say!'

We make a strange-looking threesome as we walk the short distance from our house into town; Eddie's vice-like grip links us and pulls us forward in jerky movements as he skips towards the sounds and lights of the markets. A group of girls are up on stage, singing slightly off-key as they try out their best Beyoncé dance moves. I recognise them from my school, Year 8, I think. The sequins on their skimpy lycra outfits sparkle as they bump and grind to the beat.

'Good grief.' Mum shakes her head and clicks her tongue in her annoying old lady way. 'I wonder what their parents think about that.' As we walk around the edge of the crowd we spot the parents standing in a group in front of the stage, clapping and cheering along with the music.

Eddie's smile is on high beam. I haven't seen him this happy for months. We wander among the stalls, stopping to look at anything that catches our eye. At the lucky dip stall he clutches the two dollars I gave him earlier and rummages deep in the blue-covered prize box. When he pulls out and unwraps a plastic water gun, he's just ecstatic.

The main act climbs on stage and the kids swell with excitement as Christmas carols ring through the warm night air. We find a table where we sit to eat our dinner of thick German sausages, mustard and sauerkraut.

I belch loudly, much to Eddie's amusement, before signalling to Mum that I want to go for a walk.

She glances at her watch and wipes up spilt mustard and sauerkraut with her serviette. She stacks the paper plates before holding the rubbish out for me to take, with that annoying sarcastic smile of hers. 'I'll be heading home in half an hour. If I can't find you by then, make sure you're back by no later than 8; and by 8 I mean *pm*, just in case there's any confusion later on.'

I kiss Eddie on the top of his head, avoiding eye contact with Mum. 'Goodnight, munchkin, see ya tomorrow.' He's rubbing his eyes and I know if Mum doesn't leave soon she'll be carrying him all the way home.

Desperate to get away, I leave the noise and crowds and head for the small laneway I use regularly as a shortcut. I'm still close enough to hear muffled conversations echo behind me. A neon sign above the barber's shop casts a pool of red and green light across the pavement. I scratch around in my pocket, searching for my phone.

A shape moves towards me, almost too quick to register, before my body's slammed against the bricks. A wave of adrenalin surges through me. Two hands have me pinned to the wall. A face leans in, eyes blazing with quiet rage.

'Don't move,' it whispers.

I nod, too stunned to make any sound at all. The boy moves in and I'm caught by a powerful wave of aftershave and body odour. 'I know who you are. I know what you did.' Thick, dark hair falls forward into his eyes. His skin looks dark, but in the strangeness of the neon light it's too hard to see him clearly. 'I've got a message for you and your friends. If it's war you're after – we're happy to give it to you.'

Part of my brain registers the danger I'm in. A little voice tells me to run, get out of here, but any movement I make only forces me back further against the rough bricks. My head turns away from his, eyes squeezed shut to avoid the heat of his anger spilling over. The buzz of words keeps coming until, finally, his grip loosens. Anger boils through me and without thinking I wrench my hand free, ball it into a fist and swing at his face. He reacts instinctively. It's as if no time has passed between my punch and his strike back. For a moment, I feel nothing. Then pain floods the side of my face. I crumple to the ground and the guy steps away, surprised, as if he's not sure what just happened. Footsteps echo: someone's running down the lane towards us.

'What the hell's going on?' a voice asks, gasping for air.

A figure looms over me before squatting down to eye level. My head's swimming and stars spark at the edge of my vision. My hand traces a path across my throbbing cheek. I can just see his eyes.

He turns away and the world pulls into focus again.

'You *hit* her?' He grabs my attacker by the front of his shirt and pushes him away, forcing him to stagger backwards down the lane. 'You were only supposed to be passing on a message!'

My attacker rubs at his jaw. 'Man, the crazy bitch punched me.'

'You hit a girl. I knew this was a bad idea.' He pauses, shaking his head. 'Just go, before you screw things up even more.'

My attacker looks as if he isn't sure what to do, then he turns, reluctantly, jogs down the lane and is swallowed by the darkness. A strong arm lifts me to my feet. 'You'd better put ice on that; it's started to swell already.'

I wince and put my hand to my cheek again. My heart's racing and my brain's struggling to keep up with what's just happened. This boy is tall and lean, with the same dark hair and eyes as the guy who attacked me. He takes another look at my face and murmurs under his breath in a language I don't understand. He's confused, apologetic, then he turns and runs after his friend. And I'm alone again.

Hot tears prick at my eyes as I stumble down the lane. I think briefly about calling Cam. Shaking uncontrollably, my hand searches my pocket for my phone, but it comes up empty. My breath's heavy, barely holding back angry sobs and my brain fires a confusing number of suggestions, but really, I just want to go home.

FIVE

All night I've tossed and turned, my mind replaying the events in the lane on an endless loop. What was with that guy who showed up to 'rescue' me?

And then I remember he's one of *them*.

It's almost 9 when I drag myself out of bed, tired and sore. Mum and Eddie have already left and I have to be at St Mary's by 10, which leaves me very little time to think about the state of my face. No amount of makeup is going to disguise the shiny purple bruise around my eye. It's an impressive reminder of the situation I'm in.

I pull out my phone at least a dozen times, and start to text Cam about what's happened, but something holds me back. Up until now it felt like I was part of a game we were all playing, and I was enjoying my part in it. Being one of the group was by far the best thing that had happened to me in a long time. But now, every time I close my eyes, I'm back in that alley, my attacker covering me with his rage. The consequences, as Mum put it, were beginning to dawn on me.

Each shift at St Mary's begins with a visit to the director's office, where I get handed the jobs list. The door's open but I knock anyway. The director of nursing looks up from her paperwork; her eyes linger on the bruise on my face, a smirk creeping across her lips.

During my shift the oldies stare at me and whisper to each other as I pass by. But I'm too tired to care what anyone else thinks about me today. One of the old guys stops me in the corridor and tells me his name is Philip; turns out Philip likes a chat. He tells me all about the time he was a lad and copped a shiner during a fight at school. His dad whipped him black and blue as punishment. For some reason he thinks this is extremely funny, mentioning that all most young people need today is a good thrashing to set them right.

By the time I arrive at Jack's room, fatigue's starting to make my eyeballs itch. I knock and enter without waiting for an answer.

'Back again, eh?'

I half shrug, half nod, before slumping into the vinyl chair.

Jack adjusts his glasses and bends forward to get a better look at my face. 'That's an impressive shiner you've got there, girlie.'

'You mean this?' My hand moves protectively to my cheek and I look away. 'It's nothing.'

'Don't look like nothing.' For the first time since we met, Jack's really paying attention, really seeing me. 'Looks like a whole lot of trouble to me.'

I think about reminding him that he's just a sick old man who knows nothing about my life. But he looks at me with eyes that say, *Girlie, I've had my fair share of trouble.* He pulls the sheet up around his waist and smooths out the creases. 'You know, I used to box back in my younger days, even fought in the professional league, before life took me in a different direction.'

It surprises me just how interested I've become in the idea that this old guy lying in the bed was once a professional boxer and I sit a bit higher in the chair. 'What happened?'

'Trouble. Trouble happened, that's what.' He smiles and taps his own eye, while staring at the bruise around mine. 'Thank God Mick found me. Pulled me out of the gutter, convinced me to fight as one of his boxers. Best thing I ever did, working for Mick. We toured the country, setting up that big tent of his, entertaining the crowds. That was real entertainment, before the days when people started to put a telly in every room of the house.'

'So, what, you'd just rock up to a town and guys would come and fight you?'

'Pretty much. You could usually tell in the first 30 seconds if they knew what they were doing. Sometimes these lads would surprise you. But we only ever fought to our opponent's level. And when all was said and done they might be sporting a black eye, but they'd have won the respect of the other blokes in town. And back then, respect was important.

'Certainly was a sight to see, Mick standing on that platform, high above the crowd, banging on that huge drum. Boom, ba-boom, ba-boom. That drum would reach right down inside your chest. Then a big brass bell would start up and Mick would start crying out in that booming voice of his: *Roll up, roll up*. And like all good showmen he knew how to bide his time, waiting till everyone stopped to see what the fuss was about. It was like he'd hypnotised 'em. Housewives, old folk, young kids, mostly people who'd never seen a real fight in their lives, their faces all focused on Mick, hanging off his every word. He'd be filling their heads with the expectation of a fight. By the end of it all, they'd be jostling each other to get inside the tent and all the fellas in town would be lining up to have a crack at us.'

He's got a cheeky smile for an old guy and, for a moment, I imagine the blood, the cheering crowd and the sound of a punch connecting, finding its mark. 'So you'd fight people from the town, people who weren't boxers? Surely they'd end up getting hurt.'

'Girlie, back then men were men. There was none of this namby-pamby public liability crap. If a man made the choice to pull on the gloves, then he accepted whatever happened in the ring.' He nods to himself. 'Back then, people took responsibility for their actions, not like nowadays.'

I shuffle in my seat and think about the words the magistrate used when handing down my sentence. I think about the restaurant and wonder who's going to take responsibility for the war that's about to spill over into our territory.

As I approach the park there are about a dozen faces from our group that I recognise. They're scattered about, under the cool of the giant fig trees. Cam's on his usual seat; he gets up and comes over with a curious expression.

'What the hell happened to you?' he says.

'I—' Tears well up in my eyes, accompanied by a strong sense of relief. 'Last night at the markets, this guy – he said he had a message for us.'

Cam tilts my jaw so he can get a better look at the damage. 'Okay, what's the message?'

I hold my voice steady and look him square in the eye. 'He said: "The war's on."'

A murmur goes around the group and I can sense their excitement. This is exactly what they've been waiting for. 'Bloody took 'em long enough,' someone yells.

‘Rory,’ Cam says, ‘did you get a good look at these guys? Do you think you’d recognise them if you saw them again?’

‘It was dark. But,’ I shake my head and kick at the leaf litter, ‘they were—’

A breeze picks up and the branches of the giant fig sway above us. A phantom of patterns swirls around us, shadows chasing the light that holds us in the middle of a dizzy spectrum. I watch his face, hoping he’ll tell me I’m going to be fine. ‘Man, they did a good job on you. Does it hurt?’

He reaches up to my face, but I knock his hand away. I’m pissed off. And I’m not even sure why. I stalk off down the hill. I have to go; I need to be anywhere but here.

Cam catches up to me and grabs me by the arm. ‘Hey, Rory, what’s wrong? We’ve been itching for this thing to get started; isn’t this what we’ve been waiting for?’

Over his shoulder I can see the group swirling with excitement and anticipation. No, I think, this is definitely not what I’ve been waiting for.

When I arrive home that night Mum takes one look at my face and loses it. She rants about how I go looking for trouble. She calls me a trouble magnet. Like everything that happens to me is done just to annoy her. When she finally calms down she surprises me by calling me over so she can check my eye. It seems I haven't reached my limit with her; not yet, anyway.

I'm so tired that I go straight to my room before dinner and fall asleep. That's when the birds come swirling into my dreams. I am 12 years old, stepping out across the back lawn, the early morning dew cool against my feet. I balance Dad's coffee in two hands as steam swirls from the cup. He's perched on a stool in the back garage, glasses slipping down his nose, leaning forward, focused.

He notices me and smiles, carefully placing the paintbrush he's just been using onto the palette. A pair of willy-wagtails flits about on the lawn between us. Dad steps outside and tosses breadcrumbs to them. A look of surprise crosses his face and he collapses to the ground. I want to run to him, to help him, but no matter how hard I try I'm frozen, caught by an invisible hand. I can feel my lips moving, the muscles in my face burning with effort, but no sound comes out. Then the birds come in their hundreds, whirling about him in a great frenzy until his body disappears inside a cloud of black and white feathers. They lift him up, and carry him further and further away from me, until the flock is nothing but a black speck floating in an endless blue sky.

I wake breathless, caught in a twist of sweaty sheets, my heart knocking against the inside of my chest. I creep into Eddie's room, squeeze into his tiny bed and wrap my body around his. Lulled by the sound of his rhythmic breathing, I slip into a dreamless sleep.

Next morning the swelling around my eye has gone down a bit and the colour has shifted. The outside edge of the bruise is turning from deep purple to an interesting shade of green. I have my breakfast, a piece of toast and a cup of instant coffee, in my room. I'm only down to do four hours today, but working at St Mary's is a far better alternative than being at home with Mum. I'd rather clean out all the toilets there than stay here listening to her snide comments.

When I arrive at St Mary's a line of storm clouds has rolled across the sky. The contrasting dark thunderheads tower behind the surrounding green bushland, lit bright with sunlight. There's a deep rumble of thunder as a lone white bird flies across the blackness and suddenly the brilliance of the bright green trees disappears as the sun is swallowed by a wall of dark grey.

I stand under the entrance awning to St Mary's watching the sky spit out fat drops of rain. The hot smell of bitumen fills the air and I linger under the awning, enjoying the excitement brought on by a sudden summer storm. Inside, the force of the rain echoes like static from a TV as the storm rolls over us. Even the residents seem to be caught in the moment, staring out of windows, remembering fragments of their youth and marvelling at the wildness outside.

By the time I make it to the high care ward, the wind has died down and the sun is beginning to break through the clouds. A light rain filters through the sunlight and taps against the windowpane. Jack's dozing, his eyes only half open when I knock on the door. I sit on the edge of the vinyl chair and notice he's asleep again. I don't have the heart to wake him. Instead, I pull *Wuthering Heights* from my backpack and begin to read to myself, pausing to look up every now and then, distracted by the flicker of Jack's pulse beneath a coarse patch of stubble on his neck.

Outside, the sky is now a clear blue but the trees are dripping. The chair creaks as I shift my position and my bookmark slips from my lap to the floor. As I bend down to pick it up I spot an envelope that has become wedged between the wall and the wheels of the hospital bed. I pull it free. Inside is a creased black and white picture of a woman. It's one of those old-style photos, the ones with a white border. It's yellowing with age and is covered by a tiny network of creases, as if it's been crushed into a ball only to be smoothed out again later. I turn the envelope over in the palm of my hand to check for some kind of clue to the identity of the woman. The front of the envelope is addressed to Jack. The handwriting leans backwards. I'm surprised when I recognise the address, *52 O'Rourke Street.* One of my friends from primary school, a girl called Sarah, used to live in that street.

I remember constantly begging my parents for sleepovers at her place after our netball game on Saturday afternoons. I smile at the memory of staying up late, eating bucketloads of popcorn and watching scary movies.

Jack stirs. His eyes open, searching the room, taking a moment to remember where he is before picking his glasses up from the bedside table and running a hand through his thinning grey hair. He raises an eyebrow when he notices me sitting beside him. 'Back again, eh?' His voice is thin and croaky and he has to cough to clear his throat. His eyes flick to the book in my lap. 'Thought you needed a bit more practice with your reading. Schools aren't what they used to be; don't know what they teach you kids these days—'

His eyes suddenly snag on the photo balanced on the palm of my hand and his face drains of colour. He licks his lips. Little strings of saliva have formed at the edge of his mouth, and he wipes them away with his finger and thumb. 'Hell's bells, girlie, where did you get *that*?'

'It was on the floor next to your bed.'

He stares at it, opens his mouth to speak, but changes his mind and closes it again. My eyes scan the room. This photo, so far, is the only evidence that he has any life outside this place. He's obviously playing dumb. There must be a good reason he's brought this photo with him.

He takes a deep breath and stares right into me. 'Do you believe in ghosts?'

A shiver brushes down my arms as if someone has touched a nerve. Before I can answer he shakes his head and sniffs. 'You young ones are either too blind or too stupid to notice them, but they exist all right.'

I consider telling him about the birds that follow me both in and out of my dreams. I open my mouth to speak, but he ploughs over the top of me.

'There are two kinds of ghosts in this world.' His face is serious. 'The first are the restless spirits who refuse to move on. Believe you me, stumbling across a ghost isn't something you forget in a hurry.'

'What, you're telling me you've seen a *real* ghost?'

'Saw it with my own eyes. Happened when we were touring out west near Windorah; bloody hot and dry out there in summer. We'd camped just outside town at Dead Man's Crossing. Had put on a right good show that night, too; crowd so excited that we were all invited for a drink down at the local pub. Well, turns out me and Duncan were the last to head back. Duncan was a huge fellow, afraid of no one in or out of the ring, but what we saw that night made both of us so shit-scared we ran all the way back to camp."

'What did you see?' I don't even bother to hide the smirk on my face.

'Well, it was a full moon and we were following the dry creek bed back to camp. Even though it was the middle of summer, suddenly it got really cold. Then, from out of nowhere, comes the glowing figure of a young lad, walked right by us. Something I'll never forget, the look of utter desperation on that lad's face, like he was searching for something he could never find. Next morning we asked around town and found out some young bloke named Euston had drowned in the creek years back. They reckoned that over the years his grave had been submerged by floodwater and other times it had been covered over by the huge dust storms that swept through the area, but the four posts that marked his place would always reappear.' He pauses, his eyes unfocused, remembering. I can tell he's enjoying himself. This is probably the first time in a long time Jack's had a captive audience.

'Okay then, what about the second type?' He couldn't miss the scepticism in my voice and he fixes me with a glare.

'These are the ghosts of our past, the ghosts we fashion for ourselves. These buggers have more power than we'd like to believe.' He stabs his finger at the photo. 'Like *that* bloody thing; been haunting me for over 50 years.'

Puddles line the edge of the road and the sky is impossibly blue, washed clean by the rain. I check my phone. It's just after 3 and I have no desire to go home and face Mum this early in the day. O'Rourke Street is not far from here, probably only a 20-minute walk. I think about Jack's envelope and its secrets. Who was that woman and what could possibly make a tough nut like him, a former boxer, so scared? There's more to his story than he's letting on.

Thanks to the storm, the air is so heavy and humid that, by the time I turn into O'Rourke Street, I'm dripping with sweat. Up the road a bit is Sarah's house. A young mother sits with two little girls under the shade of a tree, but they're too busy setting out tiny plastic cups and saucers on a checked tablecloth to notice me. I heard some talk that Sarah and her family had moved to Victoria. Primary school feels like a distant memory and hearing that she'd moved on without me knowing gave me a small pang of disappointment in my chest. One of the girls looks up and waves. I give a little wave back and continue up the street, keeping an eye on the letterboxes till I reach number 52.

It's a post-war home, lowset, brown brick, nothing out of the ordinary. The front yard's a bit overgrown, but the letterbox has been cleared of junk mail. A couple of the free local papers are scattered across the lawn, soggy and swollen with rain. I walk up to the front door and use the old metal knocker. I'm not sure what I'm expecting, but there's no answer. I wander down the concrete driveway that stretches along the side of the house to the backyard. A large shed sits facing the back of the house and runs almost across the entire width of the yard. It's shiny and new looking and feels a bit out of place. A metal chain with a padlock keeps the doors locked.

'You're a bit early, love.'

'Sorry?' My heart's suddenly beating faster. I have no real reason to be here and desperately try to think of something to explain why I'm snooping around a stranger's yard.

An old lady's leaning over the wooden fence that skirts along the driveway. She's holding up a small fluffy dog so he can see too. 'Your friends don't usually arrive till about 4, so you might be waiting for a while.' She smiles.

I nod, not wanting to let on that I don't have a clue what she's talking about.

'Do you know when Jack's coming back?' she continues brightly. 'Personally, I think he's a little old to be traipsing all over the world. Mind you, I've never really been one for travel.' The small dog she's holding lets out a series of yaps. She waits for it to finish. 'No doubt he'll have plenty of stories to share when he does get back. Jack's always been good at keeping us entertained.' She turns away, talking good-naturedly to the dog as if it's a child who has forgotten its manners.

Things are getting weird. Why on earth would Jack's neighbour think he's off travelling the world? And why are people coming to his house when he's clearly not here? I check my phone. It's only a 40-minute wait till 4; besides, at the moment it's not like I've got anywhere better to go. I flick through my tattered copy of *Wuthering Heights*, but just end up reading the same paragraph over and over. Concentrating on the story is impossible; Jack's reaction to the photo and his talk about ghosts are playing on my mind.

The sound of voices nearby sends me scurrying around the other side of the house. A group of four guys about my age are chatting to each other and making their way slowly into the yard. They're all wearing the same blue singlets, shiny shorts and high black boots. One of them opens the lock with a key, swings both doors out wide and props them open with a rock. From where I'm standing I can just make out boxing gloves and headgear hanging from hooks along the wall inside the shed. The guy who opened the door turns to his friend. When I see his face, my body goes cold. It's the guy from Friday night, the one who attacked me.

Fear comes back in a great rush. There are four of them and I'm on their turf. I don't want to think about what'll happen if I'm discovered. I move closer to the brickwork and force myself to slow down my breathing. Muffled voices mingle with the sound of punches landing in quick succession. I crouch low, moving quietly down the side of the house, checking that the coast is clear before bolting for the safety of the street.

*

The next morning I'm getting ready for my shift at St Mary's when it dawns on me that my copy of *Wuthering Heights* is missing. A few minutes of half-arsed searching and I tip the contents of my backpack onto the bed, sorting through a squashed banana and old receipts. The last time I remember having the book was reading on Jack's back patio. I get a terrible sinking feeling in the pit of my stomach. I'm going to have to go back there to look for it.

I leave home well before my shift starts, retracing my steps, eyes searching the footpath for my book. I turn into O'Rourke Street and there's a bird call I immediately recognise. A willy-wagtail darts out from the shrubs onto the lawn in front of me. It flits about, swivelling its black fan-shaped tail from side to side. Cocking its head on an angle, the bird considers me. Dad's morning ritual of feeding the birds fills my mind and I'm lightheaded; my limbs feel as if they're turning to vapour. I close my eyes, expecting to vanish completely, but when I open them again, the bird's gone.

I keep to the shade under the trees till I reach number 52. The house looks much as it did yesterday. I pause for a moment to listen and check the shed. The metal chain and padlock are fastened shut and everything's quiet. I head straight for the back step, where I sat reading. I search round the corner of the house and breathe a sigh of relief. There, in a patch of overgrown grass, a bit worse for wear, is my book.

A dark figure rounds the corner. For the second time that morning, my heart almost stops. I'm exposed, halfway across the backyard, with nowhere to hide. Our eyes lock. I can tell by the confused expression on his face that he's trying to place me, that maybe he knows me from somewhere. Before he has time to react I take off. I don't get more than a few metres before a strong hand catches my elbow and pulls me off balance. Everything happens in slow motion. There's a smack as I hit the ground and the contents of my bag spill out across the concrete patio.

'Oh my God, what are you doing?' My voice is teetering on the edge of hysteria. 'In what universe is it okay to attack me?' I take a good look at his face and finally recognise him. He's the guy from Friday night – the one I thought was coming to help me until he turned out to be one of *them*. 'Twice!' I spit. I should be scared, but I'm not. I'm far too pissed off.

He runs a hand through his thick dark hair, squats and picks up my books, stacking them into a neat pile before sliding them into my backpack. He grabs the apple that's rolled away into the garden bed and wipes it on his shirt before he drops it in and zips it up. 'Is that all you're going to eat today – one apple? It's no wonder you're so skinny.' His speech is clear and deliberate, as if he's choosing each word carefully.

I'm watching my bag swing from his outstretched hand. *What the hell?* 'You attack me for a second time, and then you think it's okay to comment on my weight?' The boy raises his hands to protect himself, then narrows his eyes as he puts the pieces together.

'You're the girl from the markets.'

His voice has a sharp edge to it, then his eyes settle on the bruise on my cheek and slowly his face softens.

I scramble up off the ground and wince as a sharp pain shoots through my wrist. He drops the backpack, wraps his arm around my shoulders and leads me over to the only chair on the patio.

'Wait here.' His voice is not unkind. He unlocks the shed and within a few minutes brings over a roll of white tape. Before I can object he skilfully wraps the tape from below my wrist to my fingers, across the crook of my thumb and back again. While he works I notice a faint aroma of aftershave. There are crisp lines where the thick fabric of his work clothes has been ironed. When he's finished he bends down and breaks the tape with his teeth. He sits back to admire his handiwork and watches my face as I move each of my fingers.

'I think it's only a small sprain.' Before I know what he's doing he leans forward focused on my swollen eye. The look on his face is intense, as if he's forgotten to breathe.

'I'm fine.' I flinch, surprised by the brittleness in my voice.

'The lady from next door, she told me a girl came by yesterday, before training.'

I turn back and look him right in the eye, silently daring him to say it was me.

'She seemed to think you were a nice girl but I think, maybe, you are here to spy for your friends.'

'I—' My head reels. Cam might be off my Christmas card list, but I'm not about to give this guy any information he could use against us. And why on earth would Jack, the real reason I'm here, have everyone believe he's off travelling overseas? Till I get to the bottom of this I'm not telling anyone *any*thing.

'I'm not here to spy on you, I was just—' my mind races for a believable answer, scrolling through a list of safe possibilities, '—curious.' Strictly speaking, this is true.

He narrows his eyes, weighing up my answer. 'So, you're telling me you're just being a – what do you call it? A stickybeak?' And I swear he's hiding a smile.

He stands, towering above me and shakes his head. A jagged scar runs down his cheek. 'Maybe you tell me another time, so I'm not late for work, eh?' he says, pointing at the logo on his shirt pocket.

Even though I'm burning with curiosity, part of me is relieved. Just how is this guy connected to Jack?

He stops to collect a mountain bike that's leaning against the timber fence. The wheels make a quiet ticking sound as we walk out to the road. It's obvious to me that we are both trying hard to seem relaxed.

'How did you know I would come back?' I say.

'I didn't. I always stop by on my way to work, just to check that everything's okay.'

'So you know the owner?'

'Of course.' He seems surprised by my question. 'He's a good man, a good coach.'

He slows down to keep pace with me, glancing sideways every now and then, as though he's trying to solve a complex puzzle. In the distance, traffic hums as we near the corner. I turn my face away from the passing cars, hoping no one I know drives past and recognises me.

His expression changes; he's made a decision. He turns to face me, extending a hand. 'My name's Essam.'

I look from his hand to his face and think about the look of apology on his face that night in the ally and again just now, and I tentatively hold out my hand. 'I'm Rory.'

There's strength in his grip. His hands are rough and callused.

'I train by myself most mornings, maybe you could come by? I think you could use some help with your right hook.' There's that expectant, raised eyebrow half smile again, as though he's sharing a joke.

Is he for real? Should I be annoyed or flattered by his invitation? I squint, shielding my eyes from the sun as he mounts his bike and rides off towards town.

SIX

Thursday afternoon and Eddie's watching TV while I'm sitting at my desk thinking about Cam. Arranged across my desk are photos I've dug out from my photo box. I pick up the one of Cam's tenth birthday party. He and I are sitting tight up against one another, with our arms round one another's neck. On the table in front of us is the blue and white iced racing car cake that Ellen made especially for Cam's birthday. Real Matchbox cars race round a road made of licorice straps. I remember now my mum taking the photo so Ellen could light the candles. My dad's standing behind us, his arms wrapped around both of us kids. He's pulling a goofy face for the camera. I'm looking straight into the lens, but Cam isn't. His eyes are on my dad and they shine with love.

Sometimes I remember feeling jealous of the way Cam would command his attention. After Dad died I was so caught up in my own loss that I didn't notice Cam. For a while he was still a regular visitor to our house, pestering

Mum for sleepovers. She didn't mean to brush him off, but with work, a new baby and grieving for her lost husband she just didn't have the time for him. Cam had always needed to feel a part of our family. I realise now that it wasn't just us who lost Dad; Cam lost the only man who'd stepped up to be a father to him.

Eddie makes an appearance and crawls up onto my lap. He wraps his arms around me and hangs on like a limpet. The Cam problem will have to wait.

'Wanna play hide-and-seek?' I say.

He nods solemnly. Hide-and-seek is serious business for a four-year-old.

'I go hide and you find me, okay? But first you have to count up to,' he places his little finger to his lips, mimicking Mum when she's thinking, 'one billion!' He claps his hands excitedly before dashing out of the room.

I count to one billion in millions, as loudly as I can, swivelling on my chair to pass the time.

'One billion! Ready-or-not-here-I-come!'

All the usual hiding places are checked first: under his bed, behind the curtains in the lounge, the laundry basket. There's only one other possibility left, the spare room, which, in the past few years, seems to have morphed into a junk room. Most of the space is taken up by a cupboard, a relic from Dad's share-house days, and the top's stacked high with bits and pieces of junk. I creep into the room as quietly as I can and jerk open the doors. Inside is so packed full of plastic storage boxes and old winter jackets there's no space that could possibly hide a small child.

A plastic bag tumbles out onto the floor and shoved in at an odd angle is my Year 7 art folder. Tucked in among all this junk, it's no wonder I haven't seen it for so long. I slide the folder out, careful not to dislodge the piles of stuff packed tightly around it, and clear a space on the floor. Something about the act of drawing, of looking closely, means it's possible to remember exactly where and when I drew each of the sketches.

I turn to one of the last pages in the folio and there's the portrait I'd drawn of Dad at work in his garage. He'd laughed when I dragged a camping chair out from the garage and set up on the lawn. The expression on his face is serene; an artist at work. For a long time I'm lost in the memory of Dad until Eddie stomps down the hallway, his eyes murderous, the little veins in his neck bulging. 'I'm never playing with you ever again. Ever, ever, ever!'

SEVEN

It's Friday morning and I've spent the night tossing and turning, trying to decide if I'll let Essam teach me how to box. But was the invitation for real? The idea of meeting him gives me a secret thrill. I can't stop thinking about the moment he stepped in to help me in the alley.

From beneath my curtain, the grey edge of the sky is only just lifting and kookaburras echo off in the distance. Outside, the morning air is cool and I wonder at the strangeness of being the only person awake while the rest of the world lies sleeping. It's good to enjoy the magic of a new day.

On the footpath in front of Jack's small brick house I pause and peer down the driveway at the shed. The doors are wide open and Essam is skipping on the spot, his face a mask of concentration. Our eyes make contact and he gets all tangled in the rope. He shakes his head a bit and turns away. I also do my best impression that me being here is the most normal thing in the world and adjust the strap of my backpack.

Inside the shed a huge industrial fan whirs overhead and the room reeks of stale sweat. The walls are lined with photos of young boxers and there's Essam in one of them. Windows on all sides of the shed are open, but heat still radiates from the corrugated iron as the sun begins to move higher into the sky. Essam's face gleams with a thin sheen of sweat and large brown eyes shine with satisfaction.

'Rory, hello, I didn't think you would come.' He folds the rope neatly in half and drapes it over the back of a chair. 'Have you ever done any boxing?'

I drop my backpack in the corner next to a plastic bin and shake my head. Last Friday night was the first time I'd ever thrown a punch at an actual person; not that I was about to admit that to Essam. He heads over to a metal cupboard and comes back with what look like two thick bandages. 'Maybe we should start with wrapping your hands so you can get an idea of what it feels like.' I hold out my left hand and he hooks a small loop at the end of the strap around my thumb and then begins wrapping the slightly elastic band around my wrist a number of times before moving the band up to my knuckles.

He's standing close, leaning forward and I have to resist the urge to take a step back. I'm starting to doubt my decision to come here and I focus instead on the rhythmic movement of the band around my thumb, back over the wrist a few more times until it's anchored down with a velcro tab. I flex my fingers, enjoying the sensation; my hand feels supported, strong. The process starts again on my other hand. When he's finished, Essam positions himself behind me, reaching his long arms around mine, and grabs my wrists. We are so close I can feel the warmth of his body. It's so intimate and I shake off the sudden urge to push my body back against his.

'Remember, in boxing you need to keep your hands up at all times to protect your face. Then move your arms like this.' He grips my wrist and then, with sudden force, jabs my right hand out in front of me, followed by the left.

I stifle a giggle.

'What's so funny?' He frowns slightly.

'I'm sorry, it's just – I feel stupid.'

He lets go and moves around to face me, deadly serious. 'There is *nothing* stupid about learning to protect yourself!'

'Just who is it that I'm supposed to be protecting myself against?' I ask, surprised at his sudden outburst. 'Let's not forget that you and your friend are the reason I've got this.' I jab a hand at the bruise that is now, thankfully, more of a yellow-green smudge around my eye.

'That was not supposed to happen. I am sorry, and so is Malik.'

His words sound sincere, but I'm not prepared to forgive anyone yet. What happened scared me more than I'm ready to let on. The straps on my hands take forever to unwind, then he moves directly in front of me, plants his feet so that his toes are touching the pile of spaghetti straps on the ground between us and holds me gently by the shoulders.

'Look, I didn't mean to upset you, it is just, I know—' His voice catches and he turns away. 'I just know how important it is to be able to defend yourself.'

He's standing too close; my heart's beating too fast.

'I shouldn't be here. I need to go.' I shrug free of him, scoop up my backpack and run from the shed.

I kick my backpack under the vinyl chair next to Jack's bed. The chair's got a high back and long timber armrests that scream retro. I'm starting to get quite attached to it.

'Catherine and Heathcliff can wait till later,' I say. There's no way I can handle reading about love at that level of intensity right now.

I want to come right out and ask about the boxing shed, but that would mean admitting to snooping around his house. Instead, I go for the subtle approach.

'Did you ever work as a trainer, you know, like mentoring young boxers?'

Jack looks uncomfortable, as if he's been caught with his hand in the lolly jar. 'Mentoring?' He considers the word for a moment, as if he's hearing it for the first time. 'Sure, over the years I've helped plenty of kids who've come from real bad situations. Mick did it for me, and it felt only right for me to do the same for others, to help these kids when they couldn't help themselves. That's where I stepped in.' He shifts up higher in his bed, grins and shakes his head at the memory. 'Some of them, when you first put 'em in the ring, jeez they were wild: unpredictable – dangerous, you know. I can pick that lot a mile off. That wildness, you can almost smell it on them. They're the kids trying to outrun their past. And for them, the challenge is to see what they're really made of, whether they can step up to the responsibility of coming to training, show up, work hard both in and out of the ring. The real battle is how they learn to fight for control over their demons.'

I think about Essam and the scar that winds its way down his cheek and wonder about his past.

Jack shifts his focus to me. 'That's what made you a bit hard to figure out. I've come across plenty of kids who've done bad stuff, most likely worse than you. I can pick the angry ones. But you, girlie, you don't have that wildness in you. You're feisty, I'll give you that, but not wild. Hate to say it, but when I look at you all I see in those big blue eyes of yours is someone who's finding their way. You've managed to wind up in a real mess because you're lost.'

For a moment I hold his gaze, then I snatch up my backpack and storm out of the room.

There's nothing much going on in the park apart from the deafening screech of the cicadas. A bead of sweat runs down my back and I hitch up my backpack, keeping to the shaded parts of the path. It's predicted to be the hottest December day on record and any normal person is inside with the aircon cranked up. I check Cam's rambling text and scroll through the message, wondering again what could be so urgent that he has to see me *right now*. Ever since that day when he snubbed me and my black eye so he could start planning the next move, he's been bombarding me with texts trying to get me interested in this stupid war of his. And I've ignored every single one.

Predictably, he's propped up against the trunk of our tree. He pulls out an earbud and waves me over.

'Heeey – sooo glad you could make it.'

A plastic soft drink bottle's tucked under his arm. He takes a swig of the brown liquid, tips his head to one side and gives me a dopey smile. I'm seriously surprised that Cam is drinking during the day. I've seen him occasionally drink on weekends, but this is a first.

Alcohol isn't really his thing and the sight of him like this makes me worried.

'What are you doing out here, Cam? Is everything okay?'

He grabs at my wrist and pulls me to the ground, leans forward and tucks a few loose strands of hair behind my ear.

'Sooo much better now you're here, Mizzz Rory.' He picks up the bottle and takes another swig, then offers it to me. The sweet smell of the bourbon turns my stomach and I shake my head.

'How can you drink that in this heat? You're going to make yourself so sick.'

'You sound like Gran.' He shoves the bottle at my face. 'C'mon I thou' you liked to par-tay.'

'Uh, Cam—' My voice comes out as an uncomfortable giggle. 'I've got to be at St Mary's this afternoon.' He leans across and for a moment it looks as if he's actually going to kiss me.

'What the hell are you doing?' I jerk back as my brain's trying to reconcile what's happening. Kissing Cam is the equivalent of kissing my brother. It's just wrong on so many levels. I take advantage of his confusion and pull myself up. He moves as if to follow, but he's too drunk to do it. After a few goes he manages to get to his feet. He's swaying on the spot and for a moment he looks lost, like a little boy.

'I mizz you, Rory,' he slurs. 'You're never round anymore.'

'You miss me, so you decided to stick your tongue in my mouth? Seriously, Cam, what's got into you?'

'C'mon, *carm* on.' His voice has a confused, hard edge to it. 'You're been all over me like a rash.'

'What are you talking about?' I think back desperately over the past few months and I almost can't believe what I'm hearing. What's changed between us? Is there something I've missed?

My head shakes slowly as if I'm waking up from a dream. How have all those years of friendship ended up like this?

I stumble away through the overgrown grass.

'Rory!' he yells. 'Come back!'

I rush past a middle-aged businessman talking loudly on his phone. He pauses in his conversation and I give him my best everything's-okay smile. He nods, waves and keeps on walking and it dawns on me that things between Cam and me are far from okay.

It's Saturday morning and one of the garden beds at the back of St Mary's looks like a war zone. Thin strips of leaves, stalks and flower heads lie strewn across the path. For the better part of an hour now I've been taking out my anger on the innocent weeds, imagining them as certain bits of Cam's anatomy, chopping them into tiny pieces. Feelings of shock at his behaviour have quickly morphed into anger.

And right now I'm really, really pissed off.

Just the thought of his attempt at a drunken kiss sends me into another chopping frenzy, but despite the shredded pile at my feet, the huge garden looks like it's hardly been touched. I stop to massage the cramp in my calf muscle, stretching my legs out in front of me across the concrete path.

'You'll never get rid of 'em, you know.'

I turn in the direction of the voice, and shield my eyes from the sun. One of the assistant nurses is watching me from under the small awning, coffee mug in hand. I've crossed paths with this woman several times over the past few weeks. She's the only staff member in this stinking place, apart from Fiona, who at least makes an attempt to acknowledge my existence.

'Pardon?'

'The weeds. No matter how hard you work at it, they just keep coming back.' She takes a sip of her coffee. 'Best anyone can do is just keep at it.' She drains the last of it and heads inside. The screen door snaps behind her.

The garden bed is so big that there's no way I'm going to get through it in one afternoon. Sweat trickles down my cheek and I wipe at it with the back of my wrist.

No doubt there's now a streak of dirt across my face. I straighten up and start hauling weeds, fill the wheelie bin and then drop the lid shut with a satisfying thud.

I head inside to find that Fiona's finished taking Jack's blood pressure and is writing up her observations on his chart. He's trying his best to look relaxed, but I can tell by his expression that he's been waiting for me to arrive.

'Back again, girlie? You seem to be making a habit out of these visits.'

'Jack, you know very well that I don't have anything better to do with my time.'

He gives Fiona a sly look and jerks his thumb in my direction. 'Hasn't come clean. Won't tell me what she's in for.'

'Hey, I'm not getting involved; I just work here.' Fiona backs out of the room smiling, her hands held up in surrender.

'Well, that makes us even, then. You haven't told me what you're in for, either!' I give him a friendly punch on the arm.

'That is private information. On a need-to-know basis,' he says, tapping the side of his nose. 'And you don't need to know.'

Even though Jack is clean-shaven and looks as though he's recently had a shower, the bags under his eyes seem darker. There is a hollowness to his face that I haven't noticed before. 'I suppose you're going to bore me stupid with that book of yours?'

'Yep, unless you've got something else hiding around here you'd like me to read?' I look around the room and shrug. He ignores the comment and closes his eyes.

'Jack, are you up for this today? I can just read to myself, if you want.'

'Truth be told, the sound of your voice is quite soothing – even if you do insist on reading from *that* book,' he says with a wave of his hand.

I don't know what to say to make it better, so I fill the silence with someone else's words, my voice rising and falling as I read with as much expression as I can muster. But I don't get very far. Jack's the only person I know who talks about ghosts without turning it into some kind of joke.

'Jack?'

His eyes are open; he looks dazed. The reading spell has been broken.

'Remember the other day, when you were talking about ghosts?' I'm not sure how to continue.

'Spit it out, girlie! It's not like we've got all day.' His voice is gruff, but then he smiles and winks and I can tell he's trying his best to lighten the mood.

'Well, I've been having these dreams and, well, it feels like the thing I'm dreaming about is following me, like, while I'm awake – and it's really freaking me out.'

My mouth's gone dry and my words are tumbling out on top of each other. I know I'm babbling, but I can't seem to stop myself. He reaches out and pats my hand.

His voice is soothing and encouraging. 'Tell me about these nightmares.'

'Birds – well one type of bird, really.' I bite my bottom lip hard. This all sounds so silly.

He gives me a sly look. 'An albatross?' He's making some sort of joke that's completely lost on me. 'Sorry, lass, don't mind me.' He waves it away with his hand. 'Go on.'

'It's a willy-wagtail.' I study the scuff marks on my shoes.

Jack throws back his head and laughs. Full-on belly laughing. Here I am trying to be serious and *he* thinks it's hilarious.

'What's so funny?' The hurt in my voice is obvious and his face softens.

'I've watched willy-wagtails defend their nests from birds twice their size. Feisty little things: don't think they know how to back down from a fight.' Clearly enjoying himself, he stops to tap his chin. 'Kind of reminds me of someone I know.'

Great, I remind Jack of the exact thing that's freaking me out.

'Chin up, girlie. I've always found these things have a way of working themselves out.' He looks so pleased with himself. Like he's solved the problem and now everything will be all right.

But I don't believe it.

Nothing in what he's said rings true. I can't forget the birds' cold black eyes and their silent demand for answers. Why did you leave him? they ask. Why did you leave your father to die all alone?

The next time I go to Jack's house the little dog next door is yapping at me from behind the fence. The lock on the shed is still fastened and I give it a bit of a tug, just to make sure. The metal chain makes a loud clatter as it drops back into place.

'Hello.'

I almost jump out of my skin. Essam is behind me settling his bike against the side fence. 'You are early. I like that.' His smile shows straight, wide teeth. While he rummages for his keys I check my phone. He's right: it's not quite 6.30 and I've even managed to impress myself. A month ago there would've been zero chance of me being dressed and out of the house before it was absolutely necessary. Essam already has the doors propped open and he passes me a skipping rope. This morning he's all business.

'After last week, I was sure you would not come back.'

The wooden handles feel smooth as I take hold of them. 'Last week you said how important it was for people to learn how to protect themselves.' I squeeze the rope, fighting the uncertainty in my voice. 'Well, I'd really like to learn.' It's not until the words are out that I realise the truth in them. Never again do I want to feel as helpless as that night being pushed up against the wall in the alleyway.

'How about this?' I suggest. 'Training sessions here at the shed are on neutral territory. I won't tell anyone about them if you don't.' He tilts his head to one side and nods.

We move out to the open area on the driveway. A piece of artificial turf has been laid out across the concrete. It's worn and tattered around the edges and looks like it's been liberated from the local mini-golf course. I run my foot over the rough edges and Essam shrugs.

'One of the boys brought it in to cover the cracks in the concrete.'

We turn so that the early morning sun is at our backs. I test the rope, swinging it over my head, enjoying the sound it makes as it smacks the ground.

'Before we start throwing punches we should work on your fitness. First thing you should learn is how to skip properly.'

'Doesn't everyone know how to skip?' I've got this covered. Memories of skipping in the playground before the school bell come flooding back.

'This is skipping for fitness, different from what you're used to.' He gives a quick demonstration and the rope blurs and makes a whooshing sound as it passes overhead. 'Remember to breathe in through your nose, otherwise you'll probably end up doing this.' He starts to make a loud wheezing and gasping through his mouth to make his point. 'Breathing through your nose will help keep you in control, okay?'

'So how many jumps are we aiming for?' Right now I'm expecting to maybe make it to 200 before falling in a heap.

'I usually skip for about 15 minutes, but today we aim for five.'

Five minutes?

I have a little practice just to get the feel of the rope, but after a few jumps the rope gets tangled. Essam assesses my technique, or lack of, before stepping forward to correct my stance. As soon as he touches me there's a little snap. His boots have been rubbing against the floor. 'Ow,' I say, rubbing my elbow, 'you zapped me!'

He looks slightly embarrassed. 'Sorry, that happens with me sometimes.' He fiddles with his watch, checking the second hand. 'Okay – go.'

We start off with a good rhythm, but every time I sneak a sideways glance at him I end up tangled in the rope. We've almost made it to 300 before I fall in a heap on the ground, gasping for breath.

'Don't let me stop you,' I say, fanning my hot sweaty face with my hand. Aerobic exercise has never been one of my strong points. He stops, leans over and gently touches my arm. Another zap cracks through the air.

'Stop *doing* that!' I laugh, rubbing at the sting.

He drops his long, muscular frame onto the ground next to me and crosses his legs, a sheepish look on his face. Here I am boiling and he hasn't even broken a sweat.

'Your eye is nearly all better.'

My hand moves protectively to my face. 'Yeah,' I say in a quiet voice, 'all better.' It's difficult to reconcile the violence of what happened that night with this Essam next to me. 'What made you come after me that night? How did you even know who I was?'

He flinches but, to his credit, he doesn't look away.

'I am friends with the owners of the Curry House. These people work so hard, but because the repairs were expensive they almost had to close their restaurant. Samar, the owner's youngest son, was with us that night at the markets, but it was Malik who recognised you from court. When he saw that you were alone—'

I was an easy target.

My fingers pick at the ragged edges of the artificial grass.

'We all knew it wasn't just you who trashed the restaurant. We know about your friends at the park. We wanted to let them know that what they did was not okay. The thing you wrote on the wall—'

I bite down hard on my bottom lip as I remember Cam spraying angry words in red across the walls of the restaurant.

I know nothing about Essam: his history, where he's from. The cicadas screech their endless summer song, rising and falling like a warning siren to brace for the hot day ahead. I catch his eye and smile.

'Come on,' he says. 'I want to demonstrate the jab while you're here and if we've got time I'll show you the jab-cross-hook combination.' He gets up and offers me his hand; this time there's no static shock.

EIGHT

It's Christmas day and the storm that had set in for most of the morning has finally cleared. Eddie has been onto me for ages to teach him how to ride a bike and since he got one for Christmas, now seems like the perfect time. But he looks ridiculous. He's wearing my old shin and elbow pads from the time I was into skating, topped with my black bike helmet. He refuses to wear his new bike helmet because Bob the Builder is for babies. But I'm beginning to think all the extra padding is more of a safety hazard. His tiny hand reaches out for mine and, once again, I feel an invisible tug.

We walk side by side, me pushing the small bike with my free hand, and keep to the shade along the edge of the footpath. I give his little hand a squeeze. It's sweaty, but I'm reluctant to let it go. Up until now I've not wanted to share the park with anyone, not even with Eddie. This park and my friends have always been private, like a little box I could take out and admire. And, until recently, I wanted to keep it that way.

Heading towards the park, we cross the road and I breathe a sigh of relief. Thanks to the storm the park's empty.

The grass is still wet but the afternoon sun has come back so fiercely that, apart from a few dark patches, the concrete is mostly dry. The bike path loops lazily through the park and skirts under the huge fig trees. In the middle is a flat concrete slab painted with handball and hopscotch squares that have almost faded away. Eddie wheels his bike to the flattest part of the path and looks up at me expectantly. I help him on, one hand gripping the underneath of the seat, controlling him and keeping him upright. His determination is really impressive but after a solid half hour of him falling off the bike, his patience wears thin. The helmet slips forward over his eyes and he rips it off and tosses it, so that it spins and finally stops.

'Eddie, the helmet stays on or we go home. What would Mum say if we turned up in Emergency while she was working there?'

He stomps off towards my tree and pushes himself between the giant roots that curve out from its wide trunk, clasps his arms tight across his chest. 'You have to keep your helmet on,' I say.

'It doesn't fit!' He picks up little stones and bits of dirt and tosses them in my direction.

'Look, if you can't keep your cool you'll never learn how to ride.'

'It's not fair. You said it would be easy!' He huffs and kicks as I try to squeeze into the narrow space beside him. There's no way we'll both fit, so I lift him onto my lap. After a short struggle he gives up and settles in.

He wipes a chubby little hand across his face and smears it with dirt as he does. I wrap my arms around him and squeeze his round tummy hard.

'Just remember I didn't learn to ride till I was ten, so you're already *way* ahead of me!' He nods like a sage. We sit like this for a while, cradled by the tree. A sudden gust of wind sends down a light shower of water from the wet leaves above, but we're too caught in the silence and the sharp afternoon light to care. Eddie's head rests on my chest; his hair smells like sweat and the inside of a helmet. His breathing is slow and steady and I give him a little squeeze just to check he's awake.

'Hey, wanna see something?' I point up at the higher branches.

He twists his neck and stares, wide eyed, and gives a nod, then watches carefully, following my hand and foot placements as we climb. When we can't go any further I stretch down and offer him my hand.

'Bit different from our tree at home, hey?' I try not to laugh at the frustration on his face. 'Don't worry, munchkin. Like most things, you'll grow into it.'

From this height the park takes on a different, almost magical quality. A few of the puddles on the concrete past the skate bowl are bright with reflections. I wriggle across the branch close to the trunk and beckon Eddie to follow. Two names, *Aurora* and *Cameron*, have been carved deep in the trunk just below a knot hole in the tree. Some of the letters have puckered and opened slightly and some have even started to close over. I run the tips of my fingers across them.

There's no 'woz here' or love heart around the names. Even though we were only 12 when we carved our names, we knew this place was special. Through the branches I spot the concrete bench where Cam sometimes sits. He's too busy with all the other stuff now; I doubt he even remembers any of this.

I wink at Eddie, pull out a stick from my pocket and start scratching at the rough bark under my name.

'Is that E for Eddie?' he whispers.

'Yes,' I say as I carve his name into the tree. 'That's what this "E" is for.'

The next time I arrive at training, the shed door's open and Essam is sitting on the edge of a weights machine, staring into the distance. When he looks up and smiles, his face is transformed. My breath hitches. How can some people be so bright that it hurts to look at them?

'Good morning,' he says in that goofy, formal way. On anyone else it would sound so wrong.

'So, master-of-all-things-boxing, what's on for today?' I'm trying to keep my head in training mode.

'I think you need to work on your upper body strength, so today you will be learning the ancient art of the push-up.'

Before I realise what's happening, he pulls his shirt over his head in one quick move, balls it up and tosses it onto a nearby chair and I just can't seem to stop looking. I'm still staring when some random sound escapes my lips. I clap my hand over my stupid mouth and freeze. Was that worse than accidentally farting?

'I hope you don't expect me to take *my* shirt off.' By making a joke, I'm doing my best to pretend that it didn't just happen.

Essam turns his head and I swear he's smiling, but when he turns back he's all business. He's laid out one of those blue gymnastic mats across the patchy grass outside the shed. He gets down on the mat on all fours and I try to pay attention to what he's doing.

'You need to start with your hands a bit further than shoulder-width apart, feet together. Keep your back straight and your eyes need to be fixed on a spot about a metre ahead of you.' He lifts his head to check that I'm getting it. 'I think you should start with elbows out so you can work on your chest.'

I raise my eyebrows and fold my arms across my stomach. His eyes dart to my chest, but he glances away.

'Your muscles, I mean your chest and shoulder muscles,' he stammers.

'Uh-huh, that's your story, you stick to it.' And now it's me who's hiding a smile.

He clears his throat and starts the demonstration again.

'Just like with skipping, your breathing is important. Beginners should start with exhaling on the down and inhale when you come up.'

'Okay, mister, if that's what the beginners do, show me how the advanced people do it.'

He settles into a perfect plank and pushes his body up and down really fast, exhaling and making these cute *oof-cha, oof-cha* kind of sounds. My stomach flip-flops. Essam is showing off for me. Then I catch myself. Oh, my God, I'm turning into one of those short-skirt girls who hang in the park.

When the demonstration's over Essam gets up and gestures for me to have a go. I start with the easier version of a push-up, propping myself up on my knees. When I look up, his hands are on his hips.

'Um, cheating much?'

'Hey, my sports teacher reckons this way is just as good,' I say.

He drops to the ground right in front of me. Our faces are only centimetres apart. 'Rory, I know the plank version is harder, but you can do this. Keep your knees and feet together and your body straight.' He props himself up on his side and shuffles even closer, 'Remember,' he picks up one of my hands and traces a square shape across my palm, 'to push with the four corners of your palm.' His voice trails off but his finger is still tracing a slow path and I've almost stopped breathing. I edge forward, hardly aware that I'm moving. Our faces are so close, our noses almost touching.

'Yoo-hoo, Essam!' Jack's neighbour has appeared out of nowhere and is calling from across the fence. We split apart. The spell has broken.

The next morning is spent wading in a sea of bedsheets, sorting them into the industrial-sized washing machines and dryers at the back of the complex. It feels like forever before I get to see Jack. When I arrive, he's seated in the visitor's chair, my chair, gazing out the window at the garden outside. I raise my eyebrows and purse my lips, but he just ignores me. There is no alternative but to sit on the bed. The mattress is so thin the coil of metal springs digs into my bum.

I search through the contents of my bag, slide out a Mint Slice packet and hold it up in a kind of ta-da movement. Jack squints through his thick lenses and a look of genuine surprise crosses his face. My shoulders wiggle in a dance of barely contained delight.

'I overheard you telling Fiona that you liked them,' I say. 'Plus, it's kind of a late Christmas present.'

'Oh, Rory—'

For the first time ever, he has actually used my name and it comes as such a surprise, it's like someone has pressed pause on everything around us.

'You know,' I clear my throat, 'I would actually prefer it if you called me Rory, or even Aurora – just not "girlie", okay?'

There are a few awkward moments of silence and he presses his lips together, considering. Then he catches my eye and points at the packet of biscuits on the sheets. It takes some wrestling before it's ripped open and the plastic tray slides out of the wrapper.

'Watch,' I say, 'and prepare to be amazed.'

Somehow, I manage to fit an entire biscuit into my mouth. He laughs, takes one himself and tries to copy me, but it only gets halfway in before he has to start chewing. For a moment I can't quite tell if he's chewing or choking, then chunks of chocolate biscuit spray out across the room. I reach for the jug and pour a glass of water. Even eating with old people is fraught with danger.

Someone's moving outside in the hallway and I shove the contraband packet of biscuits under the sheets. Fiona sticks her head in the doorway and gives us a suspicious grin.

'Everything okay in here?'

Jack's still coughing and his eyes are all watery.

'Nothing to see here.' His smile is charming. 'Keep moving.'

He takes another gulp, raises his half-filled glass in Fiona's direction and waves her away.

She eyes us both. 'Well, if you say so. Any problems and you know where to find me.'

As soon as she's gone Jack turns to me and passes his hand in front of his face like he's casting a spell. 'You don't need to see his ID, these aren't the old geezers you're looking for. You can go about your day, keep moving, keep moving.' He's speaking in a clipped Obi-Wan Kenobi voice and I immediately catch on and clap my hands.

'I love *Star Wars*. It's one of my favourite movies!'

'Girlie—' He catches himself, and shakes his head. '*Rory*, I was once a sucker for a good western.'

I brush the crumbs off the edge of the bed and wrinkle my brow, trying to figure out what he means. 'Sorry to break it to you, old man, but the last time I looked *Star Wars* was definitely *not* a western.'

He waggles a finger in my direction. 'Yes, it bloody is!' he snaps. 'It might be set in outer space with laser cannons and the like, but it is definitely a western. Now, hurry up and hand over another one of those biscuits.'

This time I place two chocolate biscuits in his shaky palm and reach deep into my backpack. 'But wait, there's more!' I hold up a small thermos like a trophy. He puts down the half-eaten biscuit, wipes melted chocolate off his fingers and looks at me while I unscrew the lid and pour half a cup like the nurses do when they're serving morning tea in the common room. He holds it under his nose and breathes in the smell of the coffee before sipping it slowly.

'Did that nurse out there mention I hate the coffee here?'

'Are you kidding me? Seriously, Jack, everyone at St Mary's knows that.'

'It's a bloody insult.'

When he's finished he holds out the cup, jiggles it from side to side and I pour out what's left. It smells so good, next time I'm bringing the bigger thermos, regardless of how heavy it will be. We share a wicked grin. Essam, Mint Slice, *Star Wars*, good coffee: the list of things we both like keeps getting longer. I lean over and plant a quick kiss on his cheek. His skin is soft and slack. He puts a hand up to touch it and smiles sheepishly. Stuff the rules. From now on I plan to turn our reading sessions into a mini picnic.

The receptionist at St Mary's name is Margaret, but secretly I call her the Beast. She talks into the phone so loudly that anyone within a 20-metre radius has the pleasure of hearing about her ingrown toenail. This woman could easily be mistaken for one of the residents. For starters, she's old and her face is a mask of powder, two shades too light, that accentuates the deep crevices of her leathery skin. Cheeks are painted red; the dark blush giving her the look of a child's doll. At the beginning of each of my shifts the director of nursing said that I had to I hand over my phone so I can fully concentrate on the work at hand. Now, this ritual has become a showdown between us.

She takes her time before reaching deep into her top drawer, red lacquered fingernails clawing at my phone before it's tossed across the counter. She pauses briefly to admire her perfectly varnished gel tips before continuing the conversation.

I make a show of wiping the phone clean on my shirt.

'Thanks *very* much.' Sarcasm may well be the lowest form of wit, but it sure feels good.

Outside reception, I lean against the trunk of a big gum tree. The air is filled with the shrill voices of lorikeets. The birds claw at the nearby grevillea flowers, hanging upside down, their weight bending the branches in their search for nectar. As soon as I turn on my phone a text and a voicemail message come through from Cam. From the texts I'm getting I think there's a good chance he doesn't even remember what happened at the park the other day. Even though I'm still angry at him, it's not the same without him being round and I make a mental note to talk to him sometime soon.

His message says there's a meeting at the park this afternoon and I'm expected to show. Impressively, he's managed to write the entire paragraph in caps. Wow, this really is an emergency.

'Yes, what is to be done with that rebellious Rory?' I say to the birds. With a slight pang of sadness my thumb hits the delete button.

Surprisingly, there is also a voicemail message from Essam. His warm voice sounds slightly uncertain, nervous, and without thinking I dial his number.

'Hey, Essam, it's Rory. What's this about going to the beach?'

'Rory, hello.' He clears his throat. 'I was wondering if you would like to join me for a swim?'

Something in my stomach shifts; Essam has called me outside of training. Is he up to something? 'What about our agreement? I thought only the shed was neutral territory.' There's a pause on the line.

'It's just hot and then I was thinking of you and, um, I would really like to see you.'

I'm thrown by his directness. Then guilt plucks at my conscience. Am I really going to take the step from neutral territory to the *other side*? How did these lines between what seemed right and wrong get so blurred?

'Well, it is disgustingly hot,' I reason, 'um—' And just like that, the decision's made. 'You'll have to give me an hour or so. Is that okay?' I'm surprised by my sudden feeling of excitement.

'No problem. I'll meet you at the training shed at 4.30.'

I swing my backpack onto my shoulders and start jogging in the direction of home. I'm going to be pushing it if I want time to get changed and back out to Jack's house by then. When I finally make it home the house is silent. Mum's most likely picking up Eddie and getting something for dinner on the way back from work. I bend over the kitchen sink and splash water onto my face. Sweaty clothes are peeled off in a flurry as I rummage through drawers trying to locate my togs. I scrawl a note, letting Mum know I'm heading to the beach with a 'friend' and it dawns on me that this is the first time in years I've written her a note to let her know where I am. When Dad was alive, leaving handwritten notes for each other around the house was a game the three of us played. It was one of our things – until he died.

I check the hall mirror. Cut-off denim shorts and loose fitting t-shirt. Thin straps dig into my shoulders and I make a mental note to buy new togs. I'll have to run like a maniac across town if I'm going to make it.

I arrive at Jack's house a breathless mess. Essam is sitting on the low brick fence out the front. At least he has the good grace not to mention that I'm sweating like a pig. His head dips forward. There's such respect in his eyes my breath catches in the back of my throat. And then there's the trademark one-eyebrow-raised half smile.

'Hello, Rory.'

Parked on the road is the most amazing car. It's a classic old Holden. Sky blue paintwork with beautiful curves and lines, polished so that I can see my reflection. He opens the door for me, I wriggle across into the leather seat and the door closes with a solid clunk. Inside is like a furnace and the metal belt buckle is almost too hot to touch.

'Nice car,' I say, studying the heavy old-style metal seat buckle.

'It belongs to my coach. He loves it.'

He notices my frown and adds, 'It's all okay. He said I could take her out for special events while he's away. He doesn't want the battery going flat.' I look out the window and smile to myself at being given 'special event' status.

He slides into the driver's seat, reaches across and rolls down my window. He smells delicious, a musky combination of soap and aftershave. He opens the windows, working the old-fashioned winder. His hands are ingrained with a network of black lines that no amount of soap could scrub away.

Music plays on the car stereo, and we travel in an easy silence. Both windows are wound right down and my hair whips crazily about my face. The car coasts down the hill, past the crowds at Alex Beach towards Mooloolaba. I lean out the window and close my eyes, enjoying the flash of light and shade that flicks across my eyelids. By the time Essam finds a park my hair is so wild that I give up trying to sort out the tangles.

'Don't worry about it.' He laughs at my feeble attempts to tame my hair.

'With hair this big I'm sure I look like Gene Simmons – minus his crazy outfits and makeup, of course.' He shoots me a quizzical look. 'You know – from the '80s band Kiss?'

He shrugs.

'My parents went through a stage when they used to play their CDs non-stop. Sometimes Dad'd get all dressed up and do air guitar solos.' I grab at an imaginary microphone and launch into a song.

Suddenly aware of how stupid I must look, I break off and cough into my hand. 'Of course, he was a much better performer.'

Essam raises his eyebrows and grins. He reaches up and adjusts a strand of hair from around my face. 'That was very interesting. But I like your hair like this; it makes you look—' Our eyes lock before he drops his gaze. I pretend to shade my face from the harsh sun as heat creeps up my neck and I pick my way across the hot bitumen to the beach.

Long shadows are cast across a wide stretch of sand. This time of day means there's plenty of space to spread our towels. My denim shorts are shrugged off and I pull at the shoulder straps of my togs, trying impossibly to cover more of my lily-white skin. Essam peels off his shirt. Standing next to him I feel awkward, exposed. Without a word, we head for the red and yellow flags. The sand is like warm shoes.

At the shoreline I hesitate as a wave rushes and bubbles up and wraps itself around my ankles, pulling, urging me into deeper water. Taking advantage of the lull between breakers Essam splashes past, launching himself at the face of the next wave. For a moment, light reflects off it, white and blinding, as he arches then dives under the glassy crest. I blink and rub my eyes with the heel of my hand. The next wave curves in front of me and I plunge forward, pushing through it. The salt stings my eyes and up on the other side I have to work to stay afloat. My toes reach down, probing for the bottom, and there's nothing but water. I look over at Essam through smarting eyes, happy that it's just us out here past the breakers.

Eventually we drag our wet bodies from the cool water and race each other back across the sand to our towels. We're puffed when we get there. Some way off a woman calls out and waves in our direction. She's wearing a yellow long-sleeve surf lifesaver shirt and the red and yellow cap and I check to see who she's trying to signal. As she approaches I brush the sand from my hands and shield my eyes from the low angle of the sun. Essam stands and extends a hand to the girl.

'Yasmin, it's nice to see you. I forgot you worked here.'

She laughs and shoves his arm playfully. She's very beautiful.

'Hey, work's that thing we all have to do during the week, right? This is what I do for fun.' She casts her eyes over the water. 'I love it down here.'

'And I feel much better knowing you are here to keep us all safe.' Essam shuffles his feet in the sand and nods in my direction. 'Yasmin, this is Rory. Rory: Yasmin, Malik's big sister,' he adds in a quiet voice.

Yasmin cocks her head to one side. Her eyes narrow. *I know who you are*, her eyes say. Her gaze comes to rest on my ankle and suddenly I'm exposed. In this light my tattoo is obvious, ugly. Her face clouds with anger and she grabs at Essam's arm and drags him down towards the water. And judging by her movements she's not happy.

She leaves him standing by the water and storms up to where I'm still sitting on my towel.

'You're so lucky I'm wearing this uniform. I don't know what game you're playing at.' She's towering over me, her finger jabbing at my face. Part of me wants to tell her where to go, but surprisingly a small voice in my head agrees. Essam's too good for the likes of me.

A man calls from further down the beach. She gives a sharp wave to the uniformed figure standing near the flags and I take the opportunity to hide my tattoo in the sand.

'Stay away from Essam or I will mess you up like you messed up my friend's restaurant.' She sneers at my buried foot and jogs back down to the flags.

NINE

Jack's room is empty. The nurses' station is empty, too. I hurry down the hall to the common room and breathe a small sigh of relief. There he is, tucked up in an armchair, fast asleep. There's another empty armchair in the far corner of the room. It makes a loud squealing noise as I drag it over the polished floor. Jack's eyes flutter open, a confused look on his face as he searches the room for clues to his whereabouts.

'Still here, eh?' His voice is weary and, for a moment, I'm not sure if he's talking to me or to himself. He runs a hand over his face and manages a weak smile. 'Best thing about getting old: dreams so real I'd swear I was back inside the big tent, fighting alongside my mates.'

Dust falls through the low beam of light stretched across the floor. He sleeps a lot more now. Fiona said that's what happens with the oldies near the end. In the past few weeks it's like the lines on his face have softened. Like this slow procession towards death is blurring the boundary between the real world and the world of his dreams.

His eyes seem to be fixed on those drifting specks of dust. He blinks hard and looks straight at me. 'Did I ever tell you about my Aboriginal mate, Billy? Billy the Kid, we called him. Everyone loved watching Billy fight; had the reflexes of a King Brown, that one, could duck and weave like nobody's business. Whenever it was time to pack up that big tent of ours and move on to the next town, Billy would make sure he and I rode together in the old Ford truck. No aircon back then: the windows would be wound right down, elbows getting baked by the sun.

'One day we were headed to Tewantin and we broke down right beside the town cemetery. Can you imagine? I made some stupid joke about it being the dead centre of town and that's when Billy told me about the Burial Tree that still stood in the centre of town. He reckoned it must have been hundreds of years old, this tree where his family had once buried their dead. They would wrap the body in bark and place them right between the giant roots of this tree, so eventually the tree and the body became one. When they'd visit the tree they'd pick the leaves, crush them and breathe in the scent.' Jack lifts his hands to his face as if he can smell the leaves on his hands now. He shakes himself free of the memory, lifts his glasses and wipes his eyes with the back of his sleeve. 'Imagine being able to communicate with your dead like that. You and me, we could learn a thing or two from Billy.'

I sit beside him, listening to the distant sound of the dinner cart as it rattles down the hall. In the few moments that pass, Jack's head slumps forward as he slips so easily into the world of his dreams. His eyes slide back and forth behind paper-thin lids. The edges of his lips twitch into a smile and I wonder where his dreams are taking him.

The sun has only just reached the top of the trees as I head down the driveway. Essam is out the front of the shed skipping and, judging by the sweat on his shirt, he's been at it for some time. Being early is obviously a badge of honour and has become a competition between the two of us. Impressively, he doesn't stop skipping. Instead he nods at the rope laid out for me. Let the torture begin.

This time I manage to count to 500 before grabbing my side, gasping for breath. The practice I'm doing at home really is paying off. Wheezing, I give him the thumbs-up, but step back when he tries to pat me on the shoulder.

My hand's held up in front of his face like a stop sign. 'Oi! No more with the zaps!'

He gives me one of his eyebrows-raised sideways smiles and steps back, hands held high.

'No worries.' The words roll slowly from his mouth, as if the 'o' is getting stuck on the way out and I smile at how hard he's working at sounding Australian. 'Are you ready to try the heavy bag today?'

We make our way into the shed and without waiting for an answer he selects a pair of boxing gloves hanging from the rows of hooks on the wall. They reek of sweat and leather. The gloves are tight and make my hands feel stiff and awkward. Down the back a huge cylindrical bag hangs from a metal chain bolted to the ceiling. I give it a shove and it swings back at me. It's solid and heavy: a bit like hitting a padded wall.

Essam gives it a hard punch and it swings back and forth.

'With the heavy bag you are punching, not pushing, okay? Try to think of the bag as your opponent, and use its movement to practise ducking and weaving.' He gives the bag another belt and swings his body from side to side. I copy him as I watch, bending at the knees, hands up in defensive position. 'You have to remember to only move your hands *or* your feet. So when you punch, plant your feet hard and then move your feet when you're not punching.' He moves away from the bag and I give a few tentative punches, then switch to bouncing on my toes.

'Yes, that's it. You should be on your toes like you were for our skipping drills. Now aim for only three to six punches in a combination, jab, or jab-jab-cross.'

I give it a go, trying to remember everything he's just said, arms up, move side to side, stay on my toes. The little voice in my head is on a loop. He steps forward, pulling my shoulders back and down. 'Look, you're holding your breath. Shoulders relaxed and breathe like you do when you skip: in through your nose.'

I soon find my rhythm, but have to stop and give my arms a rest as they feel like lead weights. Every now and then I glance over at him. He's using the small bag attached by what looks like a swivel hook, the speed bag. His moves are fluid and relaxed. I wonder if he even knows what it's like to lose control.

At 7.30, it's time to pack up and lock the shed. I'm feeling pretty chuffed about how the training's coming along. Essam tosses the equipment into the bins and pulls a plastic container from his bag. He peels off the lid and holds up something shaped like a small pastie.

'What's this?' I ask, taking it. It's covered in icing sugar and smells like almonds.

'It is qottab, a Persian treat.'

'Did you make them?'

'No, my auntie is about to open a café in town and she has been testing her baking on me.' He picks out the biggest piece and takes a huge bite. I give a tentative sniff and nibble at the edge.

'Mmm, this is delicious,' I say between bites, 'your

auntie's onto a winner. What are they called again?' I try to say the word and he corrects me.

'Qwo-tarrb,' I say, rolling my r's too hard and he laughs good-naturedly at my lousy pronunciation. I lick the sugar from my hands. It reminds me of Cam talking about parfaits and the awesomeness that is the Pineapple Boat.

'Essam, have you ever tried a Pineapple Boat?'

He raises his hands and shrugs, a confused expression on his face. 'A pineapple can be a boat? How is that even possible?'

I mock gasp and grab him by the hand. 'Please don't tell me you've never been to visit our iconic Big thing in town?' He thinks for a moment and shakes his head. 'How about I give you a hint, it's big and yellow and shaped like a pineapple.'

'Rory, you're not making fun of me are you?' he says in mock horror. 'No, I have never been to the Big Pineapple.'

'What, you live less than ten minutes away from this landmark and you've never been? I mean, how is *that* even possible?' I ask, pulling him over to the car. 'It's time for you to be educated in the world of Big things and there's no time like the present.'

Essam turns into my driveway so I can get changed out of my sweaty gym gear and into a fresh set of clothes. Ten minutes later when I get back to the car Eddie's gripping onto my hand, his booster seat under my other arm. Essam is leaning against the car door, waiting patiently. He gives Eddie a little wave.

'Hey, this is my little brother, Eddie.' I hold up the booster seat. 'Is it okay if he tags along? All it took were the words *Big Pineapple* and *parfaits* and – well – let's just say Eddie loves the Big Pineapple and there will be a major meltdown if I don't let him come.'

For some reason Eddie's suddenly looking all shy and hiding behind my back despite the turn he just put on in front of Mum. Essam smiles and squats down so that he is at Eddie's level.

'Would you like to show me around the Big Pineapple?'

Eddie nods and takes a tentative step forward. While I manoeuvre the booster seat into the back of the car the two boys spend the next few minutes chatting about ice-creams, baby animals and train rides.

When we arrive, Eddie pulls me up the hill from the carpark to the main entrance. A few Japanese tourists turn in our direction and smile at his enthusiasm.

He leads the way up the concrete steps and into the huge fibreglass pineapple. He gives us a blow-by-blow description of the ancient farming diorama that shows the process of growing and processing pineapples. After a few minutes he tugs on my hand again.

'Let's go all the way up to the spikey bit!' he yells, and leads us up the two flights of stairs to the very top of the building.

We step out onto the narrow walkway that leads round the top of the pineapple. The view from up here is gorgeous. Fields of pineapples are planted in neat rows that lead up to rolling hills of untouched bushland. A few ancient bunya pines grow along the edge. It is truly beautiful.

Eddie's jumping up and down on the spot, trying vainly to see over the high, solid barrier of green. Essam scoops him up and lifts him till his head clears the top of the barrier. Below us the train that loops around from the main shops to the animal farm lets out a loud, hissing toot and before we know it, Eddie's leading us back down the stairs. He grabs Essam's hand and rattles on about the animal farm and the train and if we hurry we might just catch it.

An hour later we've chugged around the fields, hand-fed kangaroos from brown paper bags filled with some sort of pellets and walked back up the steep slope to the main building. We're hot and hungry as we head through the shops crammed full of printed pineapple stuff. Tourists mill about trying to decide which pineapple printed tea towel or key ring to buy, or which flavour macadamia nut to taste.

At the entrance to the food hall, Eddie spies a coin-operated miniature yellow digger complete with a deep sand pit and I have to drag him past it and on to the ice-cream counter. I line up with Essam and count out my money, silently thankful Mum gave me some. I've only got enough for a Pineapple Boat and one parfait, though.

I look around for Eddie, wanting to ask which ice-cream he'd like and my heart leaps into my throat. Where is he? Just as I'm about to step out of the line I notice Essam already making his way through the crowd of Japanese tourists towards the toy digger we passed on the way in. Eddie's sitting on it, madly tugging and pushing at the levers, his lips pursed to make tractor noises. Essam rummages around in his pocket and places a coin in the slot. The machine levers suddenly spring to life and the look on Eddie's face is priceless. Essam turns back to me and waves.

I wait in line for my order, carry the heavy tray to the first empty table I find and place the parfait for Eddie and the Pineapple Boat for Essam on the table. The two boys suddenly appear and I hand each of them a long-handled spoon.

'Prepare yourself,' I say grandly, 'for the best pineapple ice-cream experience of your life!'

Essam can't believe the size of the half pineapple piled with chopped fruit pieces and topped with three huge scoops of ice-cream and cream.

'There is no way this is meant for one person,' he says.

'Not to worry,' I say waving a spoon of my own and sitting down next to him. 'I'm very happy to help.'

And we all dig in. Essam tips his head forward and pushes his hair off his forehead. 'Um, my auntie, the one who made the qottab? Tonight is the opening of her café and, um, I was wondering if maybe you would like to come.'

I try to smile round a mouthful of ice-cream without being too gross. Essam's trying to hide a smile too. I ignore the little warning voice in my head and search the bottom of my bag for my phone. My thumb taps out a quick text and almost immediately Essam's phone pings. He checks the message, a goofy grin on his face.

'You can pick me up at 7.' I shrug.

I pull at the hem of my dress. I can't remember the last time I wore a dress and I'm feeling like I've just spent the entire summer holidays running around in bare feet, only to have to squeeze into a pair of tight leather school shoes. It feels wrong. I study my reflection in the mirror, surprised by the curves and shape of the person who stares back at me. It's amazing what a quick trip to the local Vinnics can do. I seem to have a radar for the good stuff, but even I was amazed at my find. Who'd buy and then get rid of such a cute dress?

I turn again, tottering slightly, checking that the high heel shoes I've borrowed from Mum's cupboard actually match and wishing madly that my purple Doc Martens went with the dress instead of making me look like a fairy with attitude. Mum walks past the door and does a double-take.

'Wow.' She comes in for a better look. 'You look – beautiful.'

'Thanks,' I say, although I'm feeling a bit embarrassed.

'Stay there. I'm just going to get the camera.' She rushes off down the hall.

Eddie appears at the door shortly after. 'Ory, you smell pretty.' He plants a wild kiss on my wrist and skips out of the room, giggling. Mum fusses about with the camera and then surprises me with a hug. I pretend not to notice that her eyes are all misty. For the first time in my life, I'm actually going on a proper date and my mother is deliriously happy.

Right on 7 the doorbell rings. Before Mum even has a chance I grab my bag and hurry down to answer it. Essam is standing there, looking impossibly cute, in a cream coloured linen shirt and long pants. Before he can speak I grab his arm and push him back through the door.

'Bye, Mum. I'll be back by 10, okay?'

'Well, 9.30 would be better,' she calls through the closed door. 'Have a nice night.'

'Don't you think I should at least speak to your mother, introduce myself?' Essam says.

'Nah. She's just happy that her daughter's finally

acting like a normal 16-year-old girl. Probably best to take it one step at a time.'

He shakes his head slightly, eyes still fixed on the closed door. 'This is not the proper way to do things.'

'Trust me; it's about all she can handle for the moment. Besides, it'll give her something to look forward to, you know, for the next time you come round.' I give him a light-hearted wink as he leads me towards the sky blue Holden and I wonder again why Jack hasn't told Essam where he really is.

'The lady next door, Mrs Norton, she said your coach was away on holidays. Do you know when he's due to come home?' I'm trying to sound as if I couldn't care less.

'He said he'd be gone for about six weeks, but he wasn't sure.' Essam stops at a red light and fiddles with the air vents. 'I have a big fight in a few weeks; I'm really hoping he will be back in time for that.'

'So, what, he just left you without a coach before a big match? What's with that?' I'm surprised at Jack and more than a bit pissed off that he'd abandon Essam and the others who he's been training.

'It's not like that. He used to work at the PCYC, but when he retired a few years ago he built the training shed we've been meeting in. He lets me use it whenever I like. I have another coach who I work with at the Nambour PCYC. But to me, Jack will always be my coach.'

He parks down a narrow lane. Through the window I can see the bright multi-coloured lights someone has strung over the entrance to the Persian Tea House. All sorts of people are hanging around. Some have got drinks in their hands. Music drifts towards us. As I step out of the car I'm suddenly sure that this is a bad idea.

'After everything that's happened are you sure it's okay that I'm coming tonight?'

He leans over and places a hand on my shoulder. 'I wouldn't be bringing you if it wasn't. Besides,' he reaches around and gently peels my fingers from the car door handle, 'when they get to know you, they will like you as much as I do.'

Yeah, right. What would these people think if they knew who I really was? Essam sees the doubt on my face and starts edging me towards the doorway. We squeeze past all the hipsters crowding the entrance.

It's dark inside and it takes a few seconds for my eyes to adjust. Intricate Persian rugs cover the floor and light flickers from the coloured glass lamps that dangle from the ceiling. In one corner large floor cushions are organised around low tables. On the tables are plates full of finger food, but my eye is immediately drawn to the large canvases that line the walls.

We head for a polished wooden counter at the back of the café. A woman is pouring tea from an ornate silver pot – not like Mum does when people come round for a cuppa: this woman is *really* pouring, holding the pot high in the air so that a long gold stream of tea runs into a tiny cup. This is tea with style.

The woman places the pot on an ornate silver stand and claps her hands together. Her eyes are lined with kohl, making her face look dramatically beautiful.

'Essam, I was starting to wonder if you were coming!' she says, grabbing his face in her hands and kissing him on both cheeks.

'Seraphine, this is my friend, Rory.' I hold out my hand and she takes it firmly between her long, thin fingers. Her eyes are intense.

'Ah, our mystery girl. He speaks very highly of you.' She hands me a small gold-etched glass filled with the tea, and takes a broad sweep of the room. 'Please! Enjoy!'

We edge our way through the crowd to a quiet corner and a low set table. Settling onto the cushions, I have to fold my legs awkwardly to one side. The dress was a mistake! I tug at it and wonder if there is any way to sit on the floor in a dress and keep my dignity intact. My eyes search the faces around me. Essam is talking, but I'm only half listening. I feel vulnerable, like a fly that's been invited to have tea with a spider. He stops talking and brushes his fingers around the edges of his mouth as if he's my mirror. I shoot him a quizzical look.

'What – have I got something on my face?' I say.

'Rory,' he leans over and touches me lightly, 'we can go somewhere else if you like.'

I pull at my dress, smoothing the hem across my knees. 'No, I'm fine.'

He's just about to get up when a man wearing jeans and a check shirt, his sleeves rolled up to the elbows, wanders over and introduces himself as a workmate. After the introductions the two men settle into a conversation about, of all things, welding techniques. Essam keeps shooting me looks of support, but I don't mind being a spectator. I rest back against the wall as a steady stream of people stop by our corner to chat to him.

Seraphine has been round and topped up my tea glass three times now. Each time she does, I'm captivated by her rich, deep voice. She may only be 150 centimetres tall, but she has the attention of everyone in the room. She greets each one of them in the same gregarious manner, her brightly coloured dress flowing behind her along with the tinkle of her laughter as she moves about the room serving tea. In this crush of people, jostling each other for space, I'm amazed she can do so with such poise.

All that tea has filtered through me, though, and I'm desperate for the loo. On my way back from the toilet I slide through the crowd, stopping to study the paintings on the walls. I move slowly from each one to the next, tuning out the din of conversation. The first few pictures are of what look like the ruins of an ancient city.

The graceful lines make it easy to appreciate the artist's style. I remember watching Dad as he measured up, carefully stretching each canvas for a painting. And every time, as only Dad could do, he would lament the loss of this skill, saying most people these days wanted the speed and convenience of pre-stretched. This, he believed, weakened the soul of the painting.

'You look with the eyes of an artist.' Seraphine's dark-rimmed eyes gaze at me with an uncomfortable intensity. She stands quite close, her hands held in front of her. Now that she is no longer pouring tea, it looks as if she doesn't know what to do with them.

'They're really good.' I stammer. 'Do you know the artist?'

'They are all my work.' She motions around the room with one arm, the bangles on her wrist clattering as she moves. 'I hang them here to remind me of where I've come from, and where I am now.'

'These landscapes are of your home country?' I point to the ancient ruins. From a certain angle they look like the bleached bones of a creature that died a long time ago.

'I was born in Iran, in a town close to where Essam grew up. Iran was once ruled by the mighty Persians, a people whose culture dates back 3,000 years. You can see, even among the ruins, how magnificent the ancient city of Persepolis was. But, if you ever see this place with your own eyes, ah, it leaves you with no words.

'But, this is nothing compared to my new home.' She guides me gently through the crowd to the other side of the room. These paintings are in the same style as those on the other wall, but instead of desert haze, it is the ghostly shapes of the Glass House Mountains that loom from the canvas, shrouded in mist and mystery.

'I always believed Iran was the birthplace of civilisation. But my new home is even more ancient. You can feel it here.' She beats her chest gently with her fist. 'The Indigenous people understand this. Their connection with this land goes back around 50,000 years. This is something that should be respected, no?'

The people seem to arrive at the opening in waves and the bell on the doorway tinkles constantly. I make my way through the noise and crush of people to Essam. The man talking about welding has moved on, but been replaced by three guys. The tall one sitting closest to me smiles and my heart picks up speed. I recognise him from a photo that hangs on the wall in the boxing shed. I look to Essam for guidance, but he is staring at his shoes and running a hand through his hair. Then his expression changes like he's decided something important. He stands, licks his lips and addresses the group.

'Everyone, I would like to introduce you to my new friend. This is Rory.'

Three pairs of eyes turn to me. Thankfully, the tall boy stands and offers his hand, and I'm so relieved.

'Ah, I've been wondering where Essam has been disappearing lately. Thought he was spending all his spare time training for his upcoming match, but maybe not?' The boy smiles and winks at him. The others all laugh as Malik approaches carrying a plate piled high with food.

'Hey, Essam, have you tried the Persian baklava? De-lish.' He picks up a large diamond-shaped sticky looking cake from his plate and stuffs it in his mouth. When Malik notices me he almost chokes. If I wasn't feeling so scared right now, the expression on his face would be hilarious, but I am not laughing and neither is he. Malik chews, swallows and puts down his plate. Then his gaze settles on Essam.

'What is *she* doing here?' His voice is quietly menacing. He looks around the group for answers but none are forthcoming and his hard gaze settles back on me. 'You should not be here; you need to leave. Now.'

I bite my lip. Why did I think coming here would be okay? Why didn't we stick to neutral ground?

Essam stands and wipes his hands on a paper serviette and scrunches it up into a ball. 'Look, mate, Rory's here as my guest. I invited her.'

Malik places the plate of food on the table, his face a tangle of disbelief and confusion.

'What the hell, man? What are you even *doing* with her?'

The Indian boy sitting at the table shuffles on his cushion. 'Settle down, Malik. Why should it bother you if Essam has a date?'

'Date? You brought her as your *date*?' His voice booms in disbelief. Even over the buzz of conversation heads turn in our direction. 'Ah, now I get it.' He turns to the Indian boy, a smile curling at the edges of his lips. 'Samar, Essam hasn't told you who this girl really is. Essam, would you like to tell him, or should I?'

Samar studies me more closely and suddenly his strange Indian name falls into place. This is the boy whose family owns the Curry House; the one who was there with Malik when he spotted me at the markets.

Essam's face hardens and his hands flex by his side. 'Malik, that's *enough*!' The two boys confront each other, then Seraphine appears out of the crowd. They tower over her, but she stands her ground.

'Have you boys forgotten your manners? This is *my* café.' Her hand taps her chest and the bangles clink softly. 'And regardless of what's going on between you two, just remember that *everyone* is welcome here.'

'It's okay, Seraphine.' Essam pulls me gently to his side. 'We have to be going anyway.' He kisses her on both cheeks. We head through the packed room and out to the semi-darkness.

'I'm sorry about Malik. He needs to learn to control his temper,' he says.

'Got that message loud and clear, don't worry.' My shaking hand moves up to my face. 'What about you, though? Are you okay? I mean, aren't you and Malik mates?'

'Yeah,' he says quietly, checking his watch, dismissing my question. 'It's already 9.15. I'd better take you home.' He opens the car door and neither of us says anything. But back in the café there was a distinct shift, like the day he first introduced himself to me and invited me to box. Despite my dodgy past, Essam's sticking by me.

TEN

A week later, when I get to Jack's room he's fully dressed in long pants and a coat that looks two sizes too big for him. He's even wearing a hat with a feather tucked into the band. He looks like one of those old blokes that hang out at the TAB.

'Um, going somewhere?' I say.

'Not me, *we*. I've been cooped up in this room too long with only you and that cute nurse for company. And since I haven't been outside since they brought me into this place, *you*,' he jabs a finger at me, 'are taking *me* for a walk.' He seems to be enjoying himself. 'Thought it was about time I had a proper look around.'

I size him up as if I was a furniture removalist. 'You're not planning a breakout, are you? The director of nursing will kill me if she thinks I'm helping one of her inmates escape.'

'Yeah, inmates is right! You and I both know the only way I'm getting out of this place is feet first, in a box. Besides, I heard through the grapevine that some juvenile delinquent has been let loose in the gardens and – well – I thought it was my civic duty to check that their work was up to spec.'

'I know.' I click my tongue and roll my eyes in an exaggerated way. 'Young people today: everyone knows they can't be trusted. I'm sure the director would just love another report endorsing her theory that youth are unruly and unmanageable.'

'Too right. There's a notepad and pencil in the top drawer, better bring them along, just in case I need you to jot down some notes for my report.'

I head off in search of a wheelchair. Thankfully the nurses' station is unoccupied. Probably just as well: some new nurse is on today, most likely a temp sent out by an agency. When I arrived I worked out quickly that she didn't want to know about me.

The mid-afternoon sun is still high in the sky when I wheel Jack into the garden. It's a lot milder today, probably the coolest it's been in weeks. The sky's a clear blue with only a few of those towering, fluffy white clouds. A few of the other residents are out with visiting family members, shuffling about the pathways that wind around the complex. It's probably the busiest I've seen it. Everyone's taking advantage of the break in the hot weather. The other visiting family members watch us as we walk by. They nod and smile.

'Nice day for it,' a man calls out.

We smile back. As we pass each other on the path he stops and pats me on the arm. 'It's heartwarming to see someone your age taking an interest in their grandfather. Well done.'

Jack sticks his chest out and pulls his shoulders back, as we follow the path. I point the wheelchair towards the gazebo that's tucked in among the grevilleas. The branches are heavy with flowers and, as we approach, a handful of birds takes off from the tree, letting loose a volley of chirps and screeches. I park the wheelchair and wipe dust from the wooden bench, avoiding the bird droppings, so I can sit down beside him. It doesn't look as if this gazebo has been used in a while. Shame; the garden looks quite lovely from here.

'So, *Grandpa*, what's the verdict?'

'Not bad for an amateur.' He's smiling in that playful way of his. In fact, seeing him here in his suit and that funny hat, you wouldn't think he was so sick.

'Right.' He winks, reaches into his jacket and pulls out a packet of Camel cigarettes. 'Now for the real reason we're here.'

His hands shake and fumble with the packet so badly that I reach across and remove the corner of silver foil from the top for him. He taps the bottom of the packet with a crooked finger, pulls the long, smooth shaft out and holds it under his nose, taking in the aroma.

I'd never noticed the joints of his pointer finger were so swollen and stiff. In fact, all his fingers looked curled, as if he's still wearing gloves and his hands have been moulded into curves from years of thumping people in the ring.

I check the garden. For some strange reason I imagine the director of nursing jumping out at us with a big stick. Luckily, there's no one around.

I frown at the cigarettes. How did our roles get reversed?

'What? A man's entitled to a last cigarette.' He leans over and offers the packet, but I shake my head.

'Jack,' I whisper, 'you shouldn't be smoking. What if someone sees?'

'Bah, they can all go to hell. A man still has some rights, even in a place like this.'

I offer to light the smoke for him. The lighter's made of metal, not like those cheap plastic ones. Cam's got a Zippo, but I've never really liked it. It's got a picture of a skull with red eyes biting into a heart. Creepy as. But I can't stop looking at Jack's lighter. It's old and dinged up, worn smooth in places.

Engraved across the front are the words: *Everything happens for a reason.*

I smile and run my fingers over the engraving. It makes a zing as it lights up. Jack sucks hard on his cigarette and leans back in his chair, his eyes closed, a satisfied look on his face. When he opens his eyes again he watches me closely.

'Had that for years; it was a present from Mick when I left boxing. As a reminder, you know.'

When I hand it back he drops it into his coat pocket. 'The bit that isn't on there is that sometimes the reason is that you just make bad decisions. We all make them, you know, but it shouldn't define who we are. You get me?'

I nod. 'Um, Jack, can I ask you something?'

He blinks yes with minimal effort.

'I've got this friend who's in some trouble and I'm not sure how to help.'

'Ah, the old *T* word. Best you can do is be there for him.'

'How'd you know it was a "him"? Anyway. I've tried that, but he's not listening. In fact, he's really digging himself in deep. I'm getting worried.'

He runs his hand across his chin. 'It's true what I said about always being there for your mates. But you know Mick had a strict rule when it came to helping folks who were drowning in their problems. He was always the first to help anyone in need. To pull no-hopers like me back from the edge. He reckoned that if they slipped back into their problems a couple of times, he'd jump in and pull 'em out. But if they were stupid enough to need help a third time – well, that's when it was time to throw the towel in the ring and walk away.'

If the time comes, will I be able to make the decision to walk away from Cam?

The warmth of the sun is making me sleepy. Out of the corner of my eye I see Jack's head roll forward. He's still holding the cigarette and smoke curls from its tip, heavy with ash. The cigarette slips from his hand into his lap. My hand fumbles in his lap and he jerks awake, but it's too late. There's a small hole burnt in his pants.

'Thought you were making a move on me, girlie!' He's trying to lighten the mood but I can tell he's embarrassed.

I open my mouth to play it down, but he waves me off.

'Don't worry about me. I'm just a silly old bugger who can't even smoke a cigarette properly.'

'Come on, Jack. It's not that bad.' I give him a friendly punch on the arm. 'Afternoon tea's probably being served in the common room right now. How about I grab us some cake and make us a coffee?'

He brushes the last of the ash off his pants and tries to look cheery for me. He points at the brick building. 'Lay on, Macduff.'

The front screen door to Ellen and Cam's house is a patchwork of electrical tape, used to cover the holes where the cat has climbed up, trying to get in. I lean against the slack screen, cupping my hand around the corner of my eyes and peer into the murky darkness inside.

'Ellen, Cam?' I yell. 'Anyone home?'

Just because the timber front door is open doesn't necessarily mean anyone's at home. Both Cam and Ellen are in the habit of going out and leaving it wide open. Anytime I've pulled them up on it in the past they say that everyone knows they've got nothing valuable to steal. Deep down I think Ellen believes in the goodness of people. Me? I'm not so trusting.

Just as I'm about to call out again, there she is shuffling down the hall to see who it is. Her hair's all squished down on one side and she's pulling at her baggy cotton dress.

'Rory, love, come in.' She stifles a yawn with the back of her hand. 'Just having a bit of a nanna nap.'

The screen door creaks as I pull it open and snaps loudly back into place behind me. I give her a quick kiss on the cheek. Today she smells of sweat and the cheap perfume she only uses when she goes to town.

'Sorry to wake you, Ellen, but I really need to speak to Cam. Is he home?' I ask, peering down the darkened hallway.

She presses her lips together and shakes her head. 'That lazy lump! Haven't seen him all morning.

He doesn't listen to me, but you have my full permission to give him a wake-up call. All that boy seems to be good for at the moment is staying out late and sleeping half the day away. I can't wait for school to go back. He needs a routine to set him straight.'

She disappears into the kitchen as Cam emerges in the hallway wearing nothing but a pair of threadbare board shorts, three empty coffee mugs in his hands. 'Hey,' he mumbles in a sleepy voice, offering coffee. The memory of Cam's drunken kiss makes me shudder but he was so out of it that day, I doubt he even remembers it happened.

He makes the coffee and I head out the backyard and drag a couple of plastic chairs into the shade. At least there's a breeze out here. He appears with the drinks, but still without a shirt. I try not to stare at the Southern Cross tattoo inked on his chest. And I try not to think about the one inked on my ankle.

'Um, Cam, could you put a shirt on?'

'What – are my guns making you feel uncomfortable?' he asks as he flexes his less than impressive biceps. I pretend to laugh at his pathetic joke. I can't tell him it's the tattoo I don't want to see. He nods and slurps on his coffee, which is gross.

'So, you gunna tell me why you've been ignoring me? Why you didn't show up at the park for the meeting?' He crosses his arms and tucks his hands under his armpits.

His tone is all business, like me not being there for the meeting is more important than all the crap that's going wrong between us.

'Well, let's see, I've just been spending my entire holidays doing that little thing called *community service*.' And there's more than a hint of anger in my voice.

He shifts uncomfortably in his chair and drops his head. 'Well, we're all getting worried. Last time you were round at the park was that day you showed up with a black eye.'

Even though I'm still pissed off at him, part of me is relieved he doesn't remember his drunken kiss. I know I want to forget it.

'Look,' I say, doing my best to keep my voice even, 'you know I've been pretty bogged down with St Mary's and looking after Eddie. But I don't have a shift today, so maybe we could hang. Catch a movie?' His eyes light up and for a moment the Cam I know and love is right there, and I get this overwhelming urge to hold onto him and not let go.

He checks his watch. 'Yeah, probably could fit a movie in this arvo.'

'Great! But – we'll have to stop by St Mary's first. I just gotta pick something up from that old guy I've been reading to.'

It's stinking hot, but for once I don't mind. There's an easy feeling again between Cam and me as we walk from his house to St Mary's. Things haven't been this comfortable between us in a long time and it feels good. When we arrive at the nursing home, I bring us in through a side entrance that cuts through the laundry to the high care ward. He dawdles at the entrance, reluctant to come in.

'Hey, they're just old; they don't bite,' I scoff and grab him by the wrist. Inside, the main desk is empty and we manage to make it into Jack's room unnoticed. Jack's sitting on the visitor's chair, a sneaky little smile on his face. He's acting like he's been expecting us and I signal for him not to be so obvious. As a distraction I rush over to the side table and grab my book.

'Rory.' Jack nods in our direction. 'Was wondering when you'd be back for that.' He points to my book and then turns to Cam with an exaggerated smile. 'So, who's your friend?'

'Jack, this is Cam.'

Jack extends his hand. Cam looks unsure, but shakes it anyway.

'Rory was telling me a bit about you on the way over, 'bout you being a tent boxer.'

'And what about you?' Jack asks. 'Ever boxed before?'

'Me, box? Nah, I'm a lover, not a fighter.' He laughs.

'Not what I hear,' Jack says, his face serious.

'Oh, what have you heard?' Cam raises his eyebrows and squints at Jack.

'Your friend here is worried about you.'

'She hasn't been telling you about my fungal nail infection, has she?' Cam rolls his eyes. 'I told you, Rory, I've got it completely under control.'

'She says you're running with the wrong crowd.' Jack jerks his thumb in my direction and Cam takes a step back.

'Okay, what's going on here?' Cam's voice is half confused, half joking, like he's waiting for me to yell, *gotcha*! And Jack isn't helping. He's making weird faces at me, trying to work out how to rescue the situation, and I'm signalling for him to cut it out. Things are just going from bad to worse.

'Rory, I didn't come here for advice. I thought we were just going to the movies.' Cam steps back towards the door. 'And for your information, old man, this "wrong crowd" you're talking about,' he draws air quote marks with his fingers, 'they're my friends.' He turns on me and his tone is harsh. 'You know, the only one who's changed lately is you, Rory! Ever since you've come here you've been different. And now I know why.' He points dramatically at Jack as Fiona appears in the doorway with a clipboard tucked under her arm.

'Hey, you!' she snaps her fingers at Cam. 'This is a hospital with seriously ill patients. Either quieten down or I'm going to ask you to leave.'

'No problem, I was just on my way out.' I reach for Cam's hand but he pulls free, pushes his way past Fiona and doesn't even glance back.

I'm trying to put Cam out of my mind by throwing myself into training. Yesterday on top of our normal session Essam and I ran 3 kilometres. I woke up this morning stiff and sore and it wasn't till we were an hour into the workout that my muscles started to loosen up.

Essam holds up the boxing pads and my gloves make a satisfying thud, thud as they slap against them. I'm trying hard to focus on the drill he's just demonstrated. Jab-jab-cross – jab-jab-cross. It's like a dance; our bodies move in time to an invisible waltz. One, two, three – one, two, three, the movement is hypnotic. I focus on the red square in the middle of the pads. If I look at him I'll lose the rhythm. My arms burn, but I'm starting to enjoy the pain that comes with Essam's exercises.

I shake my arms and bounce from foot to foot. We're both puffing and shiny from sweat and it's nearly the end of our session. A light rain taps a steady beat on the shed's tin roof. I signal for a break and we both take the opportunity to grab a drink from our water bottles. I pull one glove off, tuck it under my arm and swig the water greedily.

‘Hey, is everything okay? You’ve hardly said two words all morning.’ Essam’s trying his best to sound upbeat, but he can tell something is bothering me.

‘Is it okay if we finish early today?’ I say.

We stare at each other, but after a few seconds he looks away. I shake my head, pull off my other glove and throw it into the plastic equipment bin.

‘Sure, whatever.’ He shrugs. ‘There are plenty of other things I need to do anyway.’ He stuffs his gear into his bag and I grab at his free hand.

‘Hey, there is something that I *must* do this morning. But it shouldn’t take long. If you like, we can meet up after.’

He thinks for a moment before nodding. ‘I can pick you up in a couple of hours, say 10.30? Maybe we can go for a drive.’

I try to hold back the grin that’s busting to get through, sorry that I can’t talk to him about Cam and at the same time feeling a prick of guilt over my failure to help Cam. We stare at each other for a bit longer. The rain’s still tapping above us. He links his fingers gently with mine. The warmth of his hand spreads up my arm.

‘Um, yes,’ I say. ‘That’d be great.’

A few hours later I’m on my hands and knees, cleaning the shower recess. After all the cleaning at St Mary’s, suddenly cleaning my own shower doesn’t seem so bad.

Mum comes in and stands in front of the mirror, brushing the tangles out of her wet hair. The bathroom smells like a combination of floral shampoo and lemon citrus cleaner. I stand up, stretch and wipe my forehead with the back of a gloved hand.

'You off to St Mary's?' Mum says.

'I should probably be home by 4-ish.' If she happens to think that's where I'll be all day, then it's not technically lying, right?

'That's a long shift, aren't you usually only meant to be on for four to five hours?' She takes a closer look at me. 'You're not going to the park, are you? You know I don't like the idea of you hanging around Cam anymore. That boy has been spending too much time with those thugs down there. He's changed – and not for the better.'

'No, Mum, I can honestly say that I will not be seeing Cam today.'

'You two were such good friends in primary school,' she adds wistfully. 'Shame. I really liked that boy.'

I walk past Fiona's desk and she looks up briefly from her paperwork. I'm not supposed to be here today, but she doesn't even register.

'Jack will be pleased,' she calls. I can tell from her voice that she's smiling.

Jack struggles to sit up in bed.

'What are you doing?' I say. 'For goodness sake, lie down.'

When I plant a quick kiss on his cheek it feels all soft and leathery at the same time. He looks quite pleased with himself.

'I've got a surprise for you,' I say and pull out a packet of biscuits. 'Monte Carlos!' His eyes light up and he glances at my bag with that knowing, expectant look. I pull out the bigger thermos and he gives me an approving nod.

An empty mug sits on his bedside table. It seems clean enough, but I give it a quick wipe out with the bottom of my t-shirt to be sure. I place the packet of biscuits right in his lap. He breathes in the strong aroma of the coffee and dunks one of his biscuits. I twist the sandwiched biscuits apart and lick at the cream and jam on the inside.

'What's got you in such a good mood, then? If I didn't know any better I'd say you'd met someone.' My mouth hangs in mid-lick and I stare at him. He's caught me completely off guard. It must be written all over my face because he laughs and slaps at the sheets. 'I'm right, aren't I? I knew it! Can spot it a mile off.' A wave of heat creeps up my neck. 'So do you want to tell me about him, or is it private?' Jack ploughs on without pausing. 'Always seems to be one of those two extremes with you young people. You're either fawning over each other in public, making a real spectacle of yourselves or sneaking around like it's the world's biggest secret.' He takes a closer look at me. I lace my fingers together, flexing them inside and out, and smile into my lap.

'Hmm, like that, is it? Well, I hope he's a good lad, all the same.' He pauses and gives me a sly sideways glance. 'It is a lad, isn't it?'

'Yes, Jack. It *is* a lad. You'd like him.'

You do like him, I nearly add. And right there and then I want to tell Jack everything. But I can't. The whole messy knot of him not realising that I know Essam and vice versa is best left alone. Who knows what Essam would think if he knew I had been keeping Jack's whereabouts secret from him. Besides, it's not my secret to tell and all I can manage is a big stuttery breath.

'Oh, I almost forgot,' I pull out an iPad and unzip it from the case. 'Here, put these on.' I unwind the little earbuds, gently press one of them into Jack's ear and click an icon. From the other earbud I can just make out the familiar music of the opening credits.

The introduction scrolls up the screen and Jack's eyes widen with excitement.

'The battery should last long enough for you to watch *The Empire Strikes Back* as well – if you want to.' Jack blinks several times and I can feel my eyes getting all misty, too. 'Just hit this button to start and stop.

And would you mind putting it back in the case when you're finished? I'll stop by this afternoon to pick it up. Mum'll kill me if anything happens to it.'

I bend down and give him another quick kiss on the cheek, hand him the second earbud and hit play. I pour the rest of the coffee into his cup and quietly make my way out, checking my watch. Fiona's desk is empty and I leave a couple of biscuits for her next to her in-tray.

From out in the hall I can hear Jack singing along with the music. Dum, dum, dum, da, de dum, da, de dum. The idea of him getting lost in the wild west of outer space, for at least a few hours, gives me a lovely warm feeling.

Rain taps on my umbrella as I make my way up O'Rourke Street. Out the front of Jack's house Essam is sitting in the driver's seat of the beautiful Holden. Music drifts from inside the car. Through the fogged-up windows I can see Essam, drumming his hands on the steering wheel, singing to himself. His shoulders sway to the rhythm. As soon as he notices me he jumps out into the rain to open my car door, still singing along with the music. The old seat squeaks as I slide across and drop my umbrella at my feet. Water starts to pool in the ridges of the plastic foot rest.

'So, where should we go?' I say.

'There is one place I like to visit when the weather is like this. It is a surprise.'

On the way we chat about Essam's apprenticeship as a sheet metal worker. He calls the apprenticeship his gap year.

'Why is it a gap year?'

Essam clears his throat and adjusts his grip on the steering wheel. 'I just found out I was accepted into engineering.'

'Wait, what? You're going to uni?' I adjust myself in my seat so I'm looking at the side of his face.

Essam shrugs and smiles sheepishly. 'When uni starts in February I'll drop down to part-time work.'

When he turns off the highway and takes the Landsborough road I've got a pretty good idea where he's going. The car winds its way up the steep, twisting road that heads to the Blackall Range. Up here the rain has been replaced by fog that hangs in thick grey patches. Essam takes the road to Mary Cairncross Park and not surprisingly the carpark is full. Families come up here to light fires and push sausages over hotplates. Children squeal as they run around in the big open area and kick a ball to each other over the wet grass. In the valley below the viewing platform, only the tops of three of the mountains poke through a thick blanket of fog. We wander across the road to the lookout.

'See that one there?' I point to the largest peak. 'That's Mt Tibrogargan. And that crooked one with the bent top bit, that's Mt Crookneck. Not sure what that other one's called, though.' I wander over to a stone monument. On top of it is a metal plaque that reminds me of an old fashioned sundial. Arrows radiate from it, marked with names and distances to all the major landmarks on the valley floor.

'Pity about the fog. But I guess – how do you say – we can't have it all,' he adds.

I study his face wondering what he's referring to.

'Hey,' he says, shaking me from the moment. 'Are you hungry? Maybe if we have something to eat and go for a little walk, the fog might clear.'

'Sure.' I smile. 'Sounds great.' I'm still lost in my thoughts when I head for the café and I'm surprised when his hand reaches for mine. We glance nervously at each other before crossing the road.

From our table we can just see the lookout. If the sky clears anytime soon, we'll be the first to know. Our order is brought over by an older woman. Stick thin with a tight bun of grey hair piled high on her head, she plonks down the tray and some of the coffee slops out of the cup and onto the serviette underneath. She gives me a hard stare and shuffles off.

'What's her problem?' I ask, jerking a thumb in her direction. The woman is still staring at me from behind the counter. Essam casually stirs the flower pattern into the froth on his coffee.

'I don't think it is you so much that she is angry at.' His coffee is now a mishmash of brown and cream swirls. 'My guess—' Essam leans across the table and lays an almond coloured arm next to my white lightly freckled one. The difference between us is so stark, even I am surprised.

The old woman is serving another customer now. She's all smiles and charming. Then her beady eyes flick over in our direction and her face hardens. The full weight of what's going on hits me.

'How dare she judge us!' I hiss. 'Ignorant old bitch!'

Across the table Essam at last stops playing with his coffee and tries it.

'Doesn't this bother you?' I say. 'How can you sit there looking so relaxed when she's, she's – ugh, I can't even bring myself to say what she's doing.' I shake my head. 'This is *not* okay.'

'Rory, remember how you reacted when we first met?' He's speaking softly, but the truth of his words knifes me. He's right. It wasn't so long ago that I flinched when he touched me, that I considered him one of '*them*'. She's still staring at us, her lips curled in disgust.

Did I really treat him this way? Was this my reaction?

Later when we go to the counter to pay the bill, the old woman nods to herself as she adds up the total on the register. Essam hands over the money and she stiffens, drops the change into the register and wipes her hands on her apron.

'Well,' I say loudly, making a show of holding his hand. 'Me and my *boyfriend* had better be going now.'

Her lips twitch and turn down at the corners, but she says nothing.

Outside, I giggle and punch him on the arm. He pushes my hand away and the look on his face is serious. I soon stop laughing.

Out of the corner of my eye I see the woman from the restaurant clearing our table. She's staring at us through the window so I stand on my tiptoes and plant a kiss on Essam's lips. Right now, I couldn't care less about that cow.

As Essam drives us back along the range to the tiny village of Montville I keep glancing at him sideways. It's silly, but I can't seem to stop smiling. The grey mist has closed in and there's a slight drizzle so we both hold the polka-dot umbrella above us, as we huddle together up the hill, stopping occasionally to check out the interesting string of shops. About halfway up the street there's a huge waterwheel. I'm pretty sure that it's just for show. It spins against a mossy stone wall, spilling water into a small pool below. We stop to watch and listen to the hypnotic splashing sound and it gives me an idea.

'Come on, time for a photo.' I take the umbrella from Essam and step back a bit to get him and the waterwheel in frame. He pulls a cheesy grin. 'Is that really the best you've got?' I run up beside him and slide my arm around his shoulder and relax into him. I hold my arm out and turn the camera on us. 'Selfie!' I sing.

Further up the road is an old Queenslander cottage that's been converted into an art gallery. 'There is something here I want you to see,' Essam says.

We walk up a wooden staircase onto the covered veranda. A large canvas is set out for display. It shows pointy-faced women in a row, each playing an oversized cello, committed to music of a mysterious song. Their sharp eyes follow us all the way into the shop. The woman behind the desk glances up and does a kind of double take before taking off her glasses and smiling.

'It's Essam, isn't it? How lovely to see you again.' She offers him her hand and Essam's long, lanky arms stretch on forever across the desk. 'Is Seraphine with you?'

'No, not today,' he says, shaking his head. 'Today I've brought my friend Rory. Do you mind if we have a look round?'

The woman notices me and smiles in greeting.

'You know,' she looks over at us and lowers her voice, 'out of all the painters we have on our books, Seraphine is my favourite. But don't tell my other painters I said that.'

Essam leads me through the small rooms crammed with paintings and ceramics. Across the back wall is a series of three paintings each over a metre wide and I step forward to study them more closely. The middle painting features Essam. Paint has been layered in such a way like it's been carved into the canvas, and my fingers itch to reach out and touch it. The image in the painting is of Essam the fighter: hard and fierce, different from the person before me.

'I had no idea that Seraphine did portraits. She's really good.'

'She painted that a few years ago. Seraphine has many talents, but the thing she is best at is helping people find their way in their new life. When we first arrived here she was the one who helped me feel part of the community. That is her true gift. She helps people feel at home.'

Across the wall there are other paintings by Seraphine. One is of a young girl around Eddie's age. She has the same almond coloured skin and wide brown eyes as Essam. She sits on a swing, her body tipped backwards flying up to the sky, her long dark hair streaming out behind. Bare feet stretch before her, straining with effort to climb as high as she can for the return journey. There's such a look of joy on her face. Here is a scene similar to any I've seen at the local playground. But the painting is titled *Freedom* and I find myself wondering what sort of life this little girl had before she found hers.

ELEVEN

I'm standing out the front of Jack's house. The brown brick building feels comfortable, like being with Jack himself. The shed's locked tight. It's late in the afternoon but today isn't a training day. That's probably why Jack asked me to stop by this afternoon to collect his stuff; he obviously doesn't want Essam or his neighbour to find out that he's not really on holiday.

The street seems strangely empty of cars. Even the yappy dog next door is quiet. My fingers sift through the contents of my backpack. It's so full of rubbish that I can't even see the bottom. Mental note: clean out my backpack, if for no other reason than it will smell better. Eventually, I find the house keys and the list Jack dictated to me earlier. He made it clear what I was to collect from the house, but as I've never been inside before, finding everything on the list might prove challenging.

Inside, packed away in drawers and cupboards, is this guy's whole life. The door creaks and even though there's still plenty of daylight, it's dark and gloomy. There's a strong musty smell and I push my finger under my nose to stifle a sneeze. My first thought is to open a window to get some air into the place, but there's no point going to all that trouble. I'll only be here for a minute.

Just inside the doorway is a light switch. There's a solid *clunk* and suddenly the room is filled with an orange glow. A naked glass light bulb dangles from the ceiling on a black cord. The bottom of the bulb is all black as if it's so old it's about to blow. In front of me is a chocolate coloured couch, with two armchairs next to it and a low coffee table. The television sits opposite, topped by an old-fashioned antenna with rabbit ears. Little crocheted doilies are draped across the backs of the chairs. Whoa – I didn't see that coming.

I walk through the house, peering into each room. One of the bedrooms has been set up as an office. A few trophies are scattered across a desk and papers sit neatly stacked in a pile to one side. I check the list again but my handwriting is so bad that it takes a moment to decipher it. The first item is supposed to be in the second drawer in the kitchen. Entering the kitchen is like stepping back into the '70s. The benches are a faded lime green. A white laminex table with black spindly legs is pushed up against the far wall. There's a cutting board sitting on the bench, with a hollow worn into it from years of use. Even the oven looks ancient.

Halfway out, the second drawer jams and it takes a good couple of minutes of jiggling and swearing to free it up. I tuck the blue folder that was making it jam into my backpack.

Standing tall against the far wall in the lounge, the wooden display cabinet is easy to spot. There's a black and white photo in a large metal frame. After listening to Jack's stories this scene feels familiar. A large crowd has gathered in front of a stage. Even with their backs to the camera I can tell these people are wearing their Sunday best. The men are in suits; the women wear long skirts, tweed jackets and hats. Handbags are slipped over forearms clenched tightly to their chests. In the centre a man stands on a high platform. Everyone's eyes are focused on him. Mick. He's lean and his face is lined from hard living. He's bent forward, towards the crowd, banging on a huge drum as if his life depends on it. Above his head a banner reads: *Boxers and Wrestlers, cash prizes, nobody barred.* Behind him a wooden facade is painted with images of boxers in high-waisted shorts and black boxing boots. Each one is shown in a different fighting stance. Silently my mouth moves, bringing their names to life. Archie, Lionel, Billy, Jack. Their limbs are relaxed, confident. Nothing can touch them and my mind slips away to Jack, lying in his hospital bed.

Banners and flags from the large tent in the background flap in the breeze. Standing beside Mick on the platform are half a dozen men, all wearing knee-length boxing shorts, socks pulled up high, singlets and boxing gloves. Some are wearing dressing gowns over their shorts. And there's Jack, looking relaxed, almost cocky. The man standing next to him is Aboriginal: *Billy the Kid* written across his singlet. And I smile at how comfortable they are with each other, like family. Jack's hair is thick and dark. He holds himself confidently. The eyes that peer into the camera are fierce, direct. It is Jack, but not the frail Jack I know. The metal frame is cool and heavy, and without hesitation I tuck it deep inside my backpack.

My hand digs deep into my pocket for the key he gave me to the cupboard. My fingers grasp the stiff piece of old ribbon tied to it. The ribbon looks as if it was once red but is now a dirty brown. The key turns easily but the door needs a few tugs before it swings open. Inside, the shelves are lined with piles of papers. Some are held together with rubber bands and some are in arch lever folders and I wonder what's so urgent that it needs collecting today.

My notes say the next folder is marked *Hammond and Spence*. Something about this name sounds familiar and, as soon as I see the logo, I understand why. It's the name of the solicitor's office in the centre of town. I swallow hard and pull it free from the pile. Inside is Jack's will. On the first page in a plastic sleeve is a letter. It's addressed to a Mr Stachowski, with the same street address as Jack. But who is Mr Stachowski?

I shove the stiff cupboard door shut and notice a small piece of paper that has dropped from one of the piles. It's a photo: badly creased, with a white border around the edges. Two faces look out at me. Jack's smiling face is smooth and free of worry. He's wearing a white shirt with the sleeves rolled up to the elbows, his arm wrapped around a woman whose eyes are filled with mischief. They lean towards each other, like magnets at the point of snapping together. She looks familiar and then I recognise her. She's the same woman from the photo in Jack's room. I slip it into my pocket.

One last look round the room and I hitch my backpack over my shoulder. I turn to make my way down the path when my legs go cold. Standing only a few metres in front of me is Essam. His expression is a mixture of shock and confusion.

'Rory, what are you doing?'

His eyes flick from the half-closed door to the set of keys in my hand to the bulging backpack on my shoulder. I hold the keys up but my hands shake so badly they make a soft, jingling sound.

'It's okay, Essam, I'm just getting some things for the owner.'

I can't bring myself to say Jack's name aloud, to admit that I know him, that I've known where he is all along. I'm smiling like a maniac. Shit, shit, shit.

His eyes are wide open, searching my face as if a fog has lifted and he's seeing me for the first time.

'You. Know. Jack?'

I nod, still smiling so much that my cheeks hurt.

'You never mentioned once you know Jack.' He pauses, struggling to understand what's going on. 'Not in the whole time we know each other. You know him well enough to have the keys to his house, and it never crossed your mind to tell me?' My tongue feels thick and my breath shortens. I look anywhere but at him. I want to tell him everything: how we both love Jack, how much I need him to understand the position I was put in. But my mind is blank.

'He's *not* on holidays then?' He reaches up and runs his hand through his hair as anger and frustration flash across his face. 'Why have you kept this from me?' He breathes out heavily. 'I put myself in a very difficult position for you, Rory. Even when I knew who you were, when my friends told me to stay away, I told myself they were wrong, that you were a good person who just got caught up with a bad group of people. But I see now who you really are. A liar, for a start.'

I open my mouth to speak, but there's nothing. His hands flex open and shut with anger and disappointment.

Then he turns away, walking at first. When he gets to the street, he breaks into a run.

*

By the time I arrive back at St Mary's it's getting dark. My hands have finally stopped shaking, but there's a sick heavy feeling in the pit of my stomach. Fiona looks up from behind the desk at the nurses' station and checks her watch. 'What are you doing here? I thought you'd gone home hours ago.'

'Jack asked me to collect a few things from his house.' I hold up the folder, too weary for conversation. Besides, I'm on the verge of bursting into a big ball of girlie tears, and I don't do tears. The lights are still on in his room and Jack's asleep, snoring softly. I pull the large metal framed photo out of my backpack and arrange it on his bedside table so it'll be the first thing he sees when he wakes up.

The room smells like boiled cabbage and there's a meal tray on his adjustable table, which has been pushed back at a strange angle. The smell is suffocating; I need to open the window. Immediately, the curtains billow and the room's filled with fresh cool air.

From his lonely spot on the wall Sad Jesus looks down at us and I wonder how many times he's witnessed this scene. Behind me Jack gives a sudden loud snort. The following silence is so long that even the ghosts in the curtains seem to be holding their breath. I'm just about to call for help when he starts breathing again. I fix Sad Jesus with a glare and think, *not yet*. You can't have him.

The streets are quiet, but there are signs of the living. The blue-green glow of TV sets, smells of dinner and the sound of scraping plates are comforting. The weight of the day presses in on me. I think about Essam, the look of betrayal on his face, and a little worry inside me batters its wings against my ribcage.

The Persian Tea House looks different during the day, less international chic and more cosy, like an afternoon at Grandma's house. A magazine sits open on the low table. My eyes move across the page but the words refuse to make sense. Every few minutes a tiny bell above the door rings, announcing another customer, and my heart races. Each time I find myself craning my neck to see if it's him. This feeling of helplessness about the situation I'm in is driving me crazy.

My legs have been curled to the side of me on a large cushion for too long and have started to tingle with pins and needles. As I come up onto my knees to stretch, the edge of the magazine catches my tea glass and sends it crashing to the floor. I curse and pick up the larger pieces of broken glass as Seraphine magically appears, broom and cloth in hand.

'I'm really sorry. I'll clean it up,' I say, as tea drips from the table. She gives me a sympathetic smile and hands me the dishcloth.

'I am short on help this afternoon; how about you repay me by helping out?' I start to mop up the tea. 'Besides,' she calls over her shoulder, 'if you drink any more tea *you* will start leaking all over my floor.'

A steady stream of customers keeps me busy clearing and wiping down tables: office workers, students and mothers with small children. Most of them greet Seraphine by name and chat happily like old friends. With all the empty shops in this end of town, I had wondered how the café would go. Just before closing time Seraphine takes me aside and teaches me the art of pouring tea. As she says, it's all in the wrist.

I'm busy sweeping the floor when the last customer leaves and she calls me over. 'So, are you going to tell me why you moped about my shop all afternoon?'

The angry look on Essam's face rattles around my mind, but a shrug's all I can manage. She eyes me closely before filling a cardboard takeaway box with leftover treats from the display cabinet. 'You know, when I first arrived here I had no English, no money and no idea about this country I had landed in. I didn't even know what season it would be when I got here, but what I did know, in my heart, was how blessed I was. A second chance is a precious gift, don't you think?' She folds down the lid and hands me the box. 'I do not think you should worry so. I have seen the way he looks at you. And where I come from, when a young man like Essam looks at a young woman like that, well—' She clicks her tongue. She focuses on me and cups her hand over my cheek, once again. 'I can see why Essam is so taken with you. My hope is that one day you will see it, too.'

I am finishing the dishes in the dining room at St Mary's when I notice that the chart on the wall above me has come unstuck and one of the corners has rolled in on itself. I dry my hands on a tea towel and smooth over the corner of the poster, then press it back into place.

'Oi!' It's a voice behind me. Philip, the one who likes a chat and keeps telling me that all kids these days need a good thrashing. For some reason his carry-on makes me smile. He stalks the hallways ready to talk to anyone whether they want to listen or not. 'I saw you,' he says. 'My eyes aren't that bad.'

I smile despite myself. 'And what is it, exactly, that you think you saw?'

'You.' He waggles a finger at me. 'Going above and beyond. Showing interest in things you don't have no need to.'

I wipe the bench down carefully, squeeze out the cloth and drape it over the tap to dry like Mum's tried to get me to do at home. I put on my best smile. 'Philip, I haven't got a *clue* what you're talking about.'

'Okay, lass, we'll play it your way, but just remember I've got my eye on you.' He winks before lowering himself with a grunt into the nearest chair.

One thing that surprised me when I first arrived at St Mary's was the number of chairs. You couldn't go 10 metres without coming across another cluster of them. Like circles of mushrooms after rain. What possible need was there for so many? But after a while I understood. One of the things about getting older was the need to stop for a rest, even on a short walk from the dining room to your bedroom.

'Game finished already?' I ask.

'Nah, still on the first innings.' He swats at an invisible fly and clicks his tongue. 'Bloody Poms, ahead by 200 runs *and* they've got five wickets in hand.' He shakes his head. 'Can't stand to watch. Give me a bloody heart attack.'

Over the past weeks I'd seen many of the residents make the long shuffle down to the common room to watch the game. They were usually a pretty subdued mob, but once the game started all that changed. You could hear the yelling and swearing echo down the hallways. And only a couple of weeks ago there'd been a punch-up during a match. The nurses had to be called in to break up the fight. It gives me hope to think that, despite everything that old age has taken away from them, they're still prepared to come out of their corners swinging.

Then Philip's expression shifts, like he's chewing on something sour. 'How's that mate of yours – what's his name? You know, the one you seem to have taken a shine to.'

Not much gets past Philip and I'm suddenly wary.

'You mean Jack?'

'Jack, yeah, that's the one. Nice fellow. Met him when he first arrived here a good month or so back. Watched the Aussies pummel New Zealand. Good match that.' His eyes cloud over. 'Knew I remembered him from somewhere, been driving me crazy for weeks.' He taps his head to make the point. 'Then one day it comes to me. Saw him fight once in one of those travelling boxing shows years back. Word had it he once fought in the professional league, till a car accident put a stop to all that. The papers made a big deal of it 'cause he was due to take on some Yank in a title fight of all things.' He sighs and shakes his head. 'Heard his missus died in the crash. Lord only knows how he managed to survive.'

Philip keeps talking and I keep nodding in the right places, but I'm not listening. My mind's turning. Car crash? Wife? And my mind turns to the black and white photo of Jack and the beautiful young woman.

TWELVE

I spit out a ragged bit of thumbnail and check the grey building in front of me. The library looks smaller than I remember, daggier, and for a moment I feel I've stayed away for too long. The electric doors slide open and a distant memory stirs. The room smells of mystery and adventure.

As a child, the sight of all the books lined up neatly, row after row, filled me with excitement. I imagined the books lined up as soldiers on parade, their spines standing to attention, but hiding behind faded dust-jackets were adventures, busting to get out. I'm laughing at my younger self. It's a good memory.

'Aurora Morris, you can't fool me. I'd recognise you anywhere.'

My favourite librarian, Ms Rose, her glasses perched at the end of her nose, as always, sizing up the situation. As a child I imagined she was the fairy godmother of the library. She had this weird knack for finding my next favourite book. She prided herself on knowing what each of her keen readers were about to need. She smiles at me through her steel-framed glasses. How did she get so old?

'Ms Rose, good to see you're still working here.'

She adjusts her glasses and sniffs. 'Won't be here for much longer, not after my favourite borrower up and left me without a word of explanation.' She smiles and leans on the counter. 'And now, my hubby keeps harping on about travelling around Australia in that caravan of his, said he'll go without me if I don't hurry up and retire.' Ms Rose fingers the ID tag slung around her neck and looks like she wants to say more, but changes her mind. 'So, Rory, what is it that's brought you back to us?' She gestures around the room and I know she's referring to the books. She always did talk about them as if they were alive.

'Actually, I'm looking to do a bit of research, for a school assignment,' I lie. 'I need to have a look at newspaper clippings from the 1960s.'

She raises her eyebrows. 'Very well then, but be aware that there have been some changes to the system since you were last here. It shouldn't take a smart girl like you long to figure it out.' She taps away at a computer terminal. 'You're in luck. There's a computer free, so I'll book it out for the rest of the afternoon. Try looking on Trove: it's a great search engine. If it happened in this area and it was big news it should be in *The Courier Mail.*' She smiles and drums her finger on the counter. 'Always knew we'd see you again.' I wave over my shoulder, smiling to myself, as I head for the workstation.

First I try the name Jack Sanford. Nothing. If Jack's accident was such big news, then his name should turn up something, even if it's just some news about him as a boxer. Maybe Philip's got it wrong. Maybe Jack isn't the same person as this title fighter whose wife died in a car accident.

This time I type in *fatal car crashes Brisbane City 1960s.* I scroll through pages of information till a headline catches my eye. 'Boxing Champ in Killer Smash'. This has to be it. The page opens; and with the text there's a photo. Emergency workers caught mid-stride, faces grim, rain coming down in sheets all round them. In the middle there's a tangle of metal, twisted beyond recognition.

> *Champion boxer Jack Stachowski, due to fight in a World Middle Weight title fight in two days' time, was badly injured in a head on crash on the Bruce Highway last night. A 26-year-old woman, also in the car, was pronounced dead at the scene.*

I sit back and take in this new information. Two things strike me. First, the woman who died must have been Jack's wife, Jack himself had been badly injured, and secondly, if this is my Jack, what's with the different surname?

I leave the library and head straight to St Mary's. I need to see him.

'Jack, Jack, wake up!'

He mumbles and rolls over. Reaching across to his bedside table, he fumbles for his glasses. His hand knocks them off and they clatter to the floor.

'Damn it!' His voice is thick with frustration. I bend down and scoop up his glasses in one quick motion. They're smeared with fingerprints and I give them a quick clean with the bottom of my shirt. He's got bed hair, sticking up at strange angles. He leans forward and his pyjama top gapes open, exposing a row of bony ribs. He's fading away.

'Rory,' he says in a tired voice. He runs a hand over the grey stubble that lines his cheeks and fixes me with a hard stare. 'What are you doing here? Didn't think you were supposed to be here today.'

I'm not sure where to start or even if I should be asking these questions. 'Last week, when you got me to pick up that paperwork from your home—'

'Is there a problem at the house?'

'No, everything at the house is fine, it's just—' My heart's beating faster. 'It's just that when I went to get your folders, like you asked, well, there's something that's been bothering me.'

Jack raises his eyebrows; seems like he knows what's coming.

'Who is Mr Stachowski?'

His laughter turns to a thick, heavy cough. 'Is that what's got you all hot under the collar? Not really much of a mystery.' He taps his chest. 'I am. I'm Mr Stachowski. Well, I was. I haven't gone by that name for a good long while.'

He adjusts his position, moving himself higher in the bed. He tries to pour himself a glass of water from the plastic jug but his hands shake so badly I have to guide them to stop water from spilling all over the sheets.

I think about my name, what it means to me to be named after my father's favourite painting, *Aurora*. I was seven when he first took me to the Queensland Art Gallery to see it. Not the same as seeing it in a book, he said. My name is part of my story. It's part of who I am. 'What would make you want to change your name?'

His face softens, registering my confusion. 'You have to understand that 50 years ago Australia was a very different place. Certain differences weren't tolerated well back then. My parents were just kids growing up in Poland during the first big war, saw things that kids just shouldn't see. Made the big trip in '37 and arrived here as Mr and Mrs Stachowski. Up to their dying day they always maintained that coming to Australia was the best decision they'd ever made. Just wanted to live in peace, get paid for an honest day's work, give their children opportunities they didn't have themselves. And they did. It was a good life. But it was our name and my parents' accent that marked us as different. We were treated like second-class citizens. By the time I was about 18 they'd had enough of being judged because of their name. Was a big deal for them to let go a part of who they were, but they decided it was for the best. Thought it would make them more Australian. Sanford seemed a very English name, so Sanford they became.'

'But didn't you change your name too?'

'Not at first. It was their decision to change the family name. As a kid I'd copped a few shiners behind the bike shed at school defending my name. Didn't take too kindly to being called a Polack by the other kids, but my parents were convinced it would make us more acceptable to other Australians. I was bloody born here, spoke with an Aussie accent, believed I was as Aussie as the next kid. Thought my parents were weak for worrying so much about it.

'I was pretty fiery in those days. There was no talking me around. I was 18 and felt I'd earned the right to that name. My parents were no different from those ten-pound Poms. But it always seemed to them that the white kids with the right sounding names were the ones doors opened for. Me and any other kid who was unfortunate enough to have different coloured skin, or a funny sounding name, we weren't so lucky. My father always said *giving up my name is but a small price to live with freedom.*' Jack sighs. 'But I was young and stupid, thought I knew better. Turns out my parents were right. Sometimes you need to know when to let things go and move on.'

That night I lie in bed, going over Jack's words about letting go – and wondering – when a door that has been closed inside my mind has suddenly opened and I'm hit by emotion. My hands are pressed hard against my eyes, trying to hold it in, but there's no stopping it now. It's all here, everything I've pieced together over the years of what happened, all the dark images in my mind. I can see him. The ghost of my father slumped against the frame of my bedroom door. His face strained. My head rings with his voice.

Rory. Rory. Aurora, where are you?

My hands block my ears but the voice gets through anyway. It's an echo, it's inside me, it's everywhere. He's pushed the door of my room open and I can see he's struggling for breath. His eyes are questioning, not understanding why I haven't come to his aid. If only he can catch his breath, he'll be okay. He scans the room till his eyes fix on the open window. The curtain flutters in the breeze. I've finally done what he's been expecting for some time. Snuck out when I'm not supposed to. But my timing sucks. He lays his head down on his arm as if he's just resting, half smiling through the pain.

The ghost fades from my mind and now I remember being inconsolable. Mum's trying to comfort me, saying it's not my fault. But I don't let up until she finally cracks and yells at me that I'm nothing but a drama queen, that I should stop being so selfish, that I'm not the only one who loved him.

I'd always said Mum should have insisted we all go to the hospital that afternoon for her pre-natal check-up. Then Dad would have been okay. He would have had his heart attack at the hospital and people would have been there to help him. Instead, Mum asked us both to stay behind to finish the mural of clouds and angels in the baby's room. But that was the day I snuck out to meet Cam at the park.

It wasn't Mum's fault.

It was me who left him.

The suddenness of this admission catches me and sobbing escapes in great shuddering waves.

I wake late the next morning. My eyes feel wretched and scratchy. Outside, a lawnmower hums in the distance; a child's voice is calling out somewhere down the road; as always there is the sharp chatter of the lorikeets. I'm cocooned by my sheets and, for the first time in what feels like forever, there is a seed of stillness inside.

THIRTEEN

Sneaking into St Mary's is a piece of cake. I don't seem to raise the suspicions of the residents. I'm just another vaguely familiar face in a uniform. I pull the stolen photo out of my backpack and tap Jack on the arm. It takes a few moments for him to register who I am and I enjoy watching his expression change from one of confusion to a smile that's broad and genuine.

'So what brings you to this neck of the woods?'

'This.'

I hold up the yellow envelope and, as expected, what little colour is in his cheeks drains away.

'I know about the accident.'

There's a sharp intake of breath and for a moment he might be making up some story.

'So you want to talk about ghosts.' For a moment I feel like I've done something right, that I've made things better. But his expression is anything but pleased. 'Did you know it was me driving that night? That it was my fault she died?' His voice catches and he takes a moment to compose himself.

'But – it was an accident.' My voice is small, quiet.

'Coroner's report came back saying it was just an accident, a result of the bad weather. But you want to know the truth? I wasn't paying attention to the road that night. I was just so wrapped up in *myself*, thinking about that damn title fight. About winning.' His crooked fingers are clenched as if he's trying to keep this moment from falling apart.

And all I'm left thinking is how one mistake, one error of judgement, can break a man forever.

I glance at Sad Jesus and suddenly I'm really pissed off.

'Jeez, I thought *I* was the world champion of self-blame but you leave me for dead! No matter how you spin this, the crash was an accident.' I'm talking to Jack like he's a small child who can't seem to keep up with the bloody obvious. 'Not your fault. Not anyone's fault. Period.'

The air feels heavy and he's saying nothing. Just a few silent tears running down his face. I reach out for his hand but he pulls away. The movement is small but it's enough to cause a sharp pain. The dinner cart rolls in and without ceremony a plastic tray is deposited on the adjustable table where it will sit till it goes cold.

FOURTEEN

Cam and I haven't spoken since that disastrous day in Jack's room, but I really need to speak to him now. Ellen answers the door. The moment she sees me, she looks worried.

'Rory, what are you doing here so early. Is everything all right?'

I give her a tight smile. 'Um, can I speak to Cam?'

She hesitates, then nods. 'He got in late last night. Probably a good thing you came by, otherwise he'd be in bed till lunch time.' She tries to smile but her face is a road map of worry.

I'm smart enough to know not to wake Cam without a peace offering, and I head straight to the kitchen to make a cup of coffee. His bedroom door is half open. Through the darkness comes the sound of steady breathing. His room stinks of old socks and there are clothes and books everywhere. Three empty mugs are parked on his bedside table with what used to be half a cheeseburger, still in its wrapper. The coffee is steaming in my hands.

'Cam!' Loud and sharp. He lurches upright in bed, eyes half open. Part of me gets a kick out of waking him.

'Wha— what's wrong?' He swings around, groans and face-plants, pulling the pillow over his head.

'Cam.' I poke him with my free hand. 'We've got to talk.'

He lifts his head from under the pillow and tries to open his eyes. 'What time is it?'

'It's already 9, come on!' I wave the mug at him. 'Just how you like it: double strength, no sugar.' Reluctantly, he sits up, pulling the sheet around his body in a rare display of modesty, and grabs the mug. 'Look, what are you even doing here?'

Part of me wonders that too. I remember Cam's lack of support when I faced the magistrate; Cam trying to kiss me; Cam storming out of St Mary's.

'I've been hearing talk. I've heard you guys are planning something for Australia Day,' I say. My voice is soft but menacing.

He puts the coffee down and crosses his arms over his skinny chest. 'What's your problem? We're doing this for *you,* for what they did to you. They can't smack a girl in the face and get away with it. I thought you'd be happy.'

It takes a massive effort to keep my voice steady.

'Why would I be happy that some other random kid gets their head smacked in?'

'They won't be just *some kid*; they're frigging boaties.'

'Cam! You sound just like those crazies you hang with.'

'Hey, we're not the problem here, *they* are!' he explodes. 'I'm starting to really worry about you, Rory. You never show up at the park anymore; you don't talk to me. Ever since you were jumped by those guys you've been different.'

He clears his throat and stares at the sheets crumpled up around his waist. 'You know there has to be some kind of payback for what they did to you. When you didn't show up the other week we had a vote. Everyone agreed, since you copped community service, we needed to do something on your behalf.'

His voice has gone all quiet. My mind drifts back to that night at the restaurant and I cringe now at what I was part of, how ridiculously proud we were of ourselves. I should be mad at him, storm out and make a scene, but I'm reluctant to leave. He still looks like the same old Cam, but everything about me has changed.

The next day Jack's sitting up in bed, the local paper in his lap, his thick-rimmed glasses folded on top of it; he's been waiting for me.

'So they keeping you busy?'

'Ugh, they had me cleaning out the storage cupboard, but I guess it's better than sitting in the hot sun weeding the garden.' I shudder when I think about finding a dead cockroach at the back of the cupboard, its splintery legs turned upwards. I hate cockroaches.

He picks up the paper and glasses and holds them out. 'You feel up to reading to an old bloke? Eyes aren't what they used to be.'

'That's what I'm here for.' I take my usual seat, pick up his glasses and look through the lenses. They're badly scratched and all smudged. 'It's no wonder you can't see out of these.' I breathe on them and try to wipe them clean with the bottom of my shirt, but I just seem to be making them worse. 'How long has it been since you got your script looked at?'

'Pah! I don't need to visit one of those eye doctors; got these from the chemist. Why would I want some snotty-nosed kid charging me hundreds of dollars for something I can get for ten bucks down the road?'

'You're telling me that you can't see well enough to read, but you won't get your eyes checked?'

'That's what I've got you for. Right – you know that nurse who's always flirting with me, what's her name?'

'*Fiona*.' I can't believe he's really forgotten.

'That's her. Well, *Fiona* was telling me I should have a look at this.' He hands me the paper. 'But I can't for the life of me make it out.' He's smiling like he knows something's going on.

The paper has been carefully folded so there's only one story showing. There's a small photo at the top – it's of Jack standing next to the heavy bag in his backyard shed.

'Hey,' I laugh, 'that's you, or have you got some mysterious twin I don't know about? How on earth did you get your photo in the paper?'

He grins. 'That photo was taken well before I landed in here. The journalist came around a few months ago to do an interview about an upcoming match one of my boys was competing in.'

'You mean one of the guys you've been mentoring?'

He nods. 'Essam: he's a great kid; got a future as a boxer if that's what he wants. Real smart, too. He's got the potential to do anything he sets his mind to. Shame he had such a tough time when he was younger.' He shrugs. 'But it's probably the reason he's so focused.'

'Essam, that's an interesting name.' Maybe he'll take the bait and fill in the blanks.

'Yeah, the kid's originally from Iran. When he was younger both his parents were imprisoned for taking part in a demonstration at the university where they taught. I've marched in a few demonstrations in my time, but imagine the courage it took to make a decision to march against that government. Both of them were arrested and ended up in jail. I don't think either of them were seen again.'

'That sounds awful,' I say, and my heart goes out to Essam. 'What a terrible way to lose your parents.'

'His auntie, she took him and got out of the country quick smart. From what Essam's told me they begged, borrowed and stole their way to Australia.'

'Did they come here in a boat, like, with people smugglers?'

'Yep, imagine, being that desperate that you'd put your life on the line for a chance at a new start.'

'So, how did you get to be working with people like this Essam, is it?'

'I'd been mentoring for years, working with the kids on the edge, you know, in clubs like the PCYC. That's where I met him. He was 13, tall and lean, but with arms on him that were all cables and cords. Right from the start he stood out. His English wasn't the best in the beginning; there was a lot of sign language and me giving demonstrations for him to follow. But he had this look in his eye, real determined like. We'd turn up to work and Essam would be there, waiting at the front, practising his drills before we'd even opened the doors.

'He was bright but real edgy, like a loaded spring. Good thing he found boxing. When he came to me he was real angry. But the world of boxing is all about the dance of giving and receiving pain, and that's a language a lot of these kids can understand. Word got around and his friends started showing up to training. Kids from countries I'd never even heard of. Cheeky buggers, loved to teach me all the swear words in their language. Always showing up with food if they couldn't pay for the session, or inviting me to their homes for a meal.' He smiles and pats his tummy, eyes glazed, remembering. 'Tell you what, though; I'd give my right arm for one of those pullaos right now.'

I pick up the paper and scan the story, half smiling to myself. 'It says here that you've got a boxing gym in your own backyard?' I'm doing my best to sound surprised, but really I just want him to come clean.

'Yeah, when I retired formally from the PCYC I thought it would be a good investment and a chance to keep seeing those kids. Turns out I was right. They're like family to me.'

I throw the paper on the bed. Okay, enough's enough.

'Jack, if they're like family how come they've never been to visit?' I want to hear him say it, that he's deliberately kept them in the dark all this time, that his deception means Essam may never talk to me again.

Jack's fingers play with the edge of his sheets, nervously smoothing out an imaginary crease. After a long silence he lets out a sigh. 'They don't come to visit because they don't know I'm here.'

His voice is all quiet. He looks out the window and

I pretend not to notice as he blinks and wipes his eyes with the back of his hand. 'Last thing those kids need is to be burdened with all this.' He gestures around the room and all I feel is an immense sadness. I thought hearing him admit this out loud would bring me some sort of satisfaction, but it doesn't. All I can see is how much the boys mean to him.

I turn back to the paper and start reading the article again. I check the time and date on my phone twice to be sure before holding up the paper triumphantly.

'Get your suit on, Jack. We're taking a field trip.'

It doesn't take much to convince him, but his hands are shaking so badly that I have to help him get dressed. When he puts the felt hat on, the one with the little feather sticking up, he looks ready. I have to stifle a laugh before thinking through the logistics of what I'm about to do.

'You really do look like a bookie with that hat on.'

Jack grabs my hand; his face a mix of excitement and concern. I can tell he's thinking about the director and the consequences of what we're doing 'There are some people you just shouldn't fight, no matter what. Are you sure this little excursion is going to be worth it?'

There is no question in my mind that for the first time in years I'm in control. This decision feels completely right and I'm going through with it regardless of the consequences. 'Don't sweat it, Jack, I've got this covered.'

Right now, I really don't give a damn if the director of nursing doesn't see it my way. Besides, the goal is to be back before anyone even notices we're gone. I ring for a cab and sneak down the hall to find a spare wheelchair. But first I really need to make sure Fiona doesn't end up in the shit. If this all goes pear-shaped I don't want her getting into trouble on my account. I check that the nurses' station is clear before steering the wheelchair through the heavy entrance door. What Fiona doesn't know won't hurt her.

The best thing about doing all the shit jobs around this place for the last few months is that I know exactly where all the side access doors are, and all the pathways that will get us out to the main road without being seen. The wire mesh gate squeezed between the manicured hedge swings open without protest. I've arranged for the cab to pick us up from the cul-de-sac at the side of the complex. I've seen visitors park here sometimes, but never any of the staff.

The mid-afternoon sun disappears behind a bank of clouds as the maxi-taxi pulls up. I pat my pocket, checking for my wallet, grateful that last time I looked there was over 40 dollars in my account. That should be more than enough to cover a return trip. By the time the driver gets the wheelchair in the back Jack and I are buckled up.

'Where are we off to today?' The driver's an older man, Indian maybe, with a soft accent.

'Nambour PCYC, thanks.'

Jack turns to me and laughs. He's rubbing at his chin so hard that I can hear the scratching. 'From the moment I first laid eyes on you I knew you were a smart cookie,' he says. He bangs the walking stick on the floor of the cab and gives a great big belly laugh. I resist the urge to tell him to calm down. The last thing I need is to be doing CPR on a 76-year-old man I've *borrowed* for the afternoon. Despite the risk, I can't help but smile. Seeing Jack so happy and full of life, whatever the consequences, is going to be worth it.

'I can't believe I forgot it was today.' He's still smiling and shaking his head in disbelief. 'What time's the fight start?'

'One,' I say, anxiously. The large digital clock above the rear-vision mirror says 12.50. 'We might miss the start of the first round.'

The taxi driver adjusts the rear-vision mirror and glances up at us. 'Mr Jack Sanford?'

'Yes.' Jack's obviously surprised to be recognised.

'My name is Rajesh, you coached my kids in boxing at the PCYC a few years back. Even came over to our house for dinner a few times. If I remember correctly, you really liked my wife's rogan josh.'

Jack pulls his glasses out of his pocket and puts them on, squinting at the driver, who keeps glancing up from the road and smiling back at us in the rear-vision mirror.

'My two boys were Raahi and Chakor, real skinny and shy, always being bullied at school. But you trained them, gave them confidence in themselves. You know they still talk about you and their time in the gym.'

Jack nods. 'I remember them, real smart lads. What are they up to now?'

'They're both studying at the University of Queensland. Engineering and IT.' The pride in his voice is obvious.

The cab pulls up outside the entrance to the PCYC. The carpark's chockers and people are still crowding through the doorway as I search for my cash card. As I hand it to the driver my eyes flick to the digital clock. It's two minutes past, but we might just make it in time. The match will be all over in less than 20 minutes and I ask very nicely if Rajesh would come back for us. We're on a tight schedule.

Rajesh kindly helps me manoeuvre Jack's wheelchair up the ramp before he goes back to park the cab. Inside it's standing room only. Heads turn, people see Jack's wheelchair and the crowd magically parts, allowing me to find a spot against the wall. It's a good thing Jack has his funny hat pulled down over his ears, as I'm sure someone would reognise him. Crowded into the room are people from all walks of life, some lined up against the far wall, and I recognise some of their faces from school chatting to the other guys from Jack's boxing group. The crowd is restless; it sounds as if everyone is talking at once and the tension is building. Malik recognises me from across the room and his brow furrows. His eyes travel from me to Jack and back again, like he doesn't know what to make of us. Then the ref walks into the centre of the ring and the crowd goes quiet. When Essam steps out into the centre of the ring, my heart flip-flops. He's bouncing from foot to foot, shoulders loose. Every third or fourth bounce he flicks his head to the side.

Essam is introduced first. He punches the air with a gloved fist and the crowd goes wild. I shouldn't be surprised. After all, this is his home turf, but to see so much support for Essam gives me a warm feeling. I pat Jack's shoulder and he reaches up to squeeze my hand.

The second fighter is standing in his corner quietly surveying the room. His singlet says Gympie Amateur Boxing Club. He looks tough, late 20s. His nose is flattened across his face and looks as if it's been broken in at least two places. His chin juts out; it's solid and broad and looks, quite literally, as if it could cop a lot. This guy has a rough, mad dog look about him and he's sniffing the air for his next victim. The ref announces his name and a handful of people in his corner cheer. The fighters go back to their corners. The ref, a short stout man, slaps the two opponents on the back, says something that's lost in the noise of the crowd, and then a bell sounds.

Round 1 begins in a blur of ducking and weaving. Essam's opponent is bigger than he is, but much slower. Essam is light on his feet, balanced, in control. Jack's body moves in time with the thrust and parry in the ring; his muscles contract and spasm as he silently coaches Essam through the first round. Jack leans in and points to the broken-nosed fighter.

'See how he's staying in close, trying to make it hard for Essam to punch with his longer arms? He's what they call a swarmer. Got a great chin on him, too. Good for copping the hits from being in so close.'

In between rounds the two opponents retreat to their corners. Essam stays focused, eyes down. The supporters that crowd his corner call out words of encouragement. From this angle I can see that one of the loudest supporters is a girl. She turns her head and I realise its Yasmin, Malik's sister, the one from the beach. She's even more beautiful than I remember. She keeps tucking her long shiny hair behind her ear, calling out encouragement. She's so close that in between rounds she could reach into the ring and touch him. He's not looking at her, I tell myself. But then I see him smile when she calls something in his direction. Jealousy plucks at my chest and my mouth tastes like metal.

Only two more rounds and it'll all be over.

By the third round both men are starting to fatigue and although Essam's opponent is slow, he packs a real punch. The fourth round is the decider. Essam's finding his mark, landing punches in quick succession. The bell rings and the fighters make their way, much more slowly now, back to their corners. The judges seated in front of the ring are deep in conversation. I'm pretty sure Essam has managed to get the upper hand. The other guy seemed to be on the back foot in the final round but, really, what would I know? Jack taps my arm and I lean down.

'It'll come down to points. That lad from Gympie's a solid fighter.' Some of Essam's friends from Jack's gym are pointing at us from across the room. One of them has pushed his way through the crowd and heads for us. My mouth goes dry. All I want to do is get out of here, get Jack back to St Mary's without being noticed.

The ref moves into the centre of the ring. He's standing between the fighters, dwarfed by the two opponents. Silence falls over the crowd and I breathe a sigh of relief as the boy crossing the room stands still to hear the verdict. The ref calls into a microphone and announces the winner is Peter from Gympie. Immediately, Essam's bumping his gloved hands with his opponent, congratulating him.

Just as he turns back to his corner, though, his eye catches mine. One minute the room's full of noise and movement and the next there's just us and I'm floating, drifting. In a panic I try to push Jack's chair towards the door, but between fights everyone wants to get outside for a quick bite or to stretch their legs. We're going nowhere.

Essam pulls off his gloves and steps through the ropes. His movements are so graceful, I can't take my eyes off him. He walks through a sea of people who slap him on the back, tell him he was *robbed* and *better luck next time*. And all the while his eyes are on us, shifting from Jack to me, me to Jack. I barely notice that Jack's shuffled his weight forward in his chair. He's trying to stand up. Instinctively, my arm wraps around his shoulders. I'm surprised at how his suit bunches up as if it's two sizes too big for him. His body's slightly hunched, tense; he's hypnotised by Essam's approach. His eyes are focused, but his expression gives nothing away. With a mighty effort, Jack stands as straight as he can and I let go of him. Under that frail exterior the determination of the fighter comes to the fore.

They're only a couple of feet apart now.

Essam looks uncertain, then, as Jack extends a shaking hand, his features settle. Without hesitating, Essam reaches forward. The other boys gather silently around us. They all look at Jack as though they can't quite believe their eyes.

'Nice match. I liked your counter punching in the second round.' There's a pause, Jack's head is held high, almost defiant. 'I see you're still dropping your shoulder on your follow-through.'

'You'd better talk to my coach about that,' Essam says in a quiet voice. Jack nods and stumbles back into the chair. Suddenly the boys who have been watching from the sidelines surge forward to help. Jack waves them away as if he's swatting a demented fly.

Outside it's bright and hot, but at least I can breathe again. The chair coasts down the ramp and Essam's strong hand reaches out to help me slow it down and catch it. Rajesh has seen us and is bringing the cab around and the boys crowd in, all speaking at once. Rajesh looks at the boys and then at Jack.

'Mr Sanford, look at all your wonderful students, such fine looking boys.' His head's bobbing up and down, a slightly surprised expression on his face. 'Unfortunately, I am only licensed to carry ten passengers.'

'It's okay, Rajesh,' I say, as Essam and I help Jack out of his chair and into the cab. 'There are only two for the return journey. Same place you picked us up from would be great.'

One of the younger boys standing near the back of the cab turns to his mate and whispers, loud enough for me to hear, 'What's *she* doing with Jack?'

Essam casually slides into the front passenger seat. He doesn't look back at us, but I notice Jack giving a slight nod. Essam chats to Rajesh; something about the sweet science of boxing. I send up a small prayer that I can get Jack back to his room without being discovered.

When we pull into the cul-de-sac the street is quiet. Essam insists on helping Jack from the taxi and into the chair. Jack slumps round-shouldered. He looks completely drained by the outing and I'm silently thankful that Essam's here to help.

Fiona glances up as we make our way past the nurses' station, but says nothing. Essam helps Jack back into bed, still refusing to make eye contact with me. As soon as Jack's settled he closes his eyes. I pull up the sheets, hoping Fiona has the good sense not to ask why he went to bed with his clothes on.

Essam sits in my chair, and we both watch him. I grab my bag and cast one last look in his direction. This time he looks me dead straight in the eye and there's that fluttering inside me again.

'You've been taking good care of him?'

I nod, opening my mouth to explain, but he cuts me off.

'He's going to tell me he didn't want us to worry about him.'

There's a bitter tone in his voice. He leans forward and massages his scalp roughly with his fingers. When he does speak his voice is choked with emotion. 'Jack is a great coach. I just wish he was a better friend.'

FIFTEEN

Under the awning at the front entrance of St Mary's, Essam and I stand an awkward distance apart. I scrabble through my backpack for my phone and sneak a quick glance at him. He's still in his boxing clothes and it occurs to me that he walked straight out of the ring without grabbing his gear.

'Ah, do you need to use my phone?' I hold it up in the air. 'I'm sure one of your friends would come and get you—' I break off because it feels like someone's tied a rubber band round my chest. He shakes his head. 'I don't want to speak to anyone just yet. They'll just be full of questions about Jack and, really – I don't have any answers.'

'Oh.' I tap the side of my phone in the palm of my hand, trying to read the situation and at the same time desperate for him not to leave. Not yet anyway, not before I've had a chance to explain.

'You know, the Persian Tea House isn't far from here. I'm sure Seraphine would give you a lift back to the

PCYC and, ah,' I dig out my wallet, 'maybe while you wait we could have a pot of tea? She might even give us leftovers from the glass case.'

'After today, I think we could *both* use a cup of tea.' Essam says it so seriously, I've got to laugh at how ridiculous it sounds.

'Yeah, 'cause tea drinking's *so* rock' n roll.'

He gives me his half smile and I can feel that rubber band tighten.

We walk side by side towards town and I tell him the whole story, starting with the magistrate and my community service order, filling him in on all the bits he's missed and the bits I couldn't go into. No matter how much I wanted to tell the truth about everything, I wouldn't do it at the expense of breaking Jack's trust.

We've been talking and walking for about 20 minutes and when I look up, there's the front door of the Persian Tea House.

I elbow him and open the door. 'So, you still want to live life on the edge and share a pot of tea?'

He's deep in thought before he nods and steps through. Inside, it takes a moment for my eyes to adjust. The café's empty, except for Seraphine, who's at the counter placing the leftover cakes from the glass case into cardboard boxes.

'Essam, Rory! Great to see you two.' She wipes her hands on a tea towel, trying to work out why there's some awkwardness going on between us. 'You both look as if you need tea and cake.' It's a statement of fact, like she's a paramedic admitting us to the emergency ward and tea's the only thing that'll revive us. After our conversation earlier, I can't help but let out a snort of laughter. Essam's lips curl into a suppressed grin and all it takes is one glance and we collapse into giggles.

We sit at the low table down the back of the café. He folds his legs and shifts to make himself comfortable on the cushions beside me.

'Sorry you lost your fight today,' I say.

He shrugs like he couldn't care less. 'That's not so important. I want to know more about Jack. How sick is he really?'

This is the one question I'm scared to answer, like somehow just talking about Jack dying will somehow bring it on. When I open my mouth to speak, the words get stuck in my throat. And I'm shaking and I must be crying because my face is wet with tears and Essam's right next to me, with his arms around me, but that just makes me cry all the more.

Seraphine locks up and drops us both back at St Mary's. It feels weird standing here with Essam in the late afternoon light. Even though it's a weekend, most visitors'll be gone by now.

Fiona's filling in paperwork at her desk.

'Hey, Fiona,' I say, and she jumps almost invisibly.

'Rory, you gave me a fright.' She clutches at her chest and then notices Essam.

'Fiona, this is my, ah, friend, Essam.' Fiona looks over the top of her imaginary glasses.

'Hello, Essam. I did notice you both here earlier. Can I ask what's so important that it can't wait till tomorrow?'

'We just have to see Jack again. I promise we wouldn't have come if it wasn't important.'

She sizes us up and checks her watch. 'I've got no idea what's been going on here today, or what you two are up to, *or* why Jack was in bed, fully dressed.' I open my mouth to explain, but she raises both hands as if it's a hold-up. 'In twenty minutes my shift will be finished and by that time both of you need to get out of here or else I'll end up getting the sack.'

Jack's door is open and he's propped up by pillows, dozing. On his side table are a collection of photographs: his framed boxing photo from home and two black and white photos of his wife. I'm not sure why, but they make me smile.

He opens his eyes and wipes the spit from around the edge of his mouth. I linger at the doorway and let Essam into the room. He sits in the visitor's chair.

'Essam, Rory, my two favourite people. What are you doing here?'

There's a short awkward silence before Essam starts talking about the boxing match and Jack visibly relaxes into the conversation. I give a little wave and signal to Essam that I'll leave them to their chat. This is between the two of them.

As I'm leaving, Fiona calls me over to her desk. 'Friend of yours?' she says. I'm too nervous to really smile. 'I take it he's a friend of Jack's, too—'

'No, I wouldn't say they're friends.' I'm wondering what the right word is. 'More like family.'

The next morning Mum's at the kitchen table in her dressing gown and Eddie's watching the Sunday cartoons when I come in, panting and sweating. She takes off her glasses and puts down the paper. Her eyes are wide with disbelief as I stick my head under the tap for a drink. Like a slow-motion mime she checks her watch and then looks back at me with fake surprise.

'I don't know who you are, or where you came from, but can you tell me what you've done with my daughter?'

I wipe my mouth with the back of my hand and, even though I'm gasping for breath, I manage to roll my eyes. 'Yeah, right. Very funny.'

'How far?' She's trying not to laugh and I answer just to see her reaction.

'Almost five kays, maybe,' I say, going for casual.

Mum thinks about that, then opens the paper and gives it a snap. 'Is today your last day at St Mary's?'

'Yep, only four hours of community service to go.' I drum my fingers on the bench, thinking. 'Might still drop in from time to time, though. Jack certainly won't be able to cope without someone bringing him decent coffee.' She gives me a puzzled glance but goes back to the paper.

I'm halfway up the stairs before Mum calls me back. 'Rory, I'm not proud of your arrest and community service sentence, but I *am* proud of the way you've handled yourself in the past few weeks.' She bites her lip. 'Your dad would be proud too.'

Heat pricks at my eyes and I escape out to the lounge before she notices. Eddie's glued to the TV. When he sees that my eyes are all teary he rests his little head on my shoulder and pats my arm like Mum used to. We can hear the sound of plates being stacked in the dishwasher.

'Come on, Eddie, time to get ready for playgroup.' Before he has a chance to argue I flick off the TV and bundle him upstairs.

The sky is a clear blue when I arrive for my final shift and, as the director of nursing's not in her office, I collect my final jobs list from the Beast in reception. Maybe I'm imagining things, but even she seems friendly today, probably only because, with me finishing up, it means one less responsibility for her.

As I wander past the common room Philip appears, pushing his walker. The TV screen flickers in the background and I can hear the noisy gun fight of a western. It's a classic, Philip tells me, *The Searchers*. When I say I haven't heard of it, he scoffs and tells me once again how terrible it must be to be me. But I just smile and he does too.

'Heard today was your last day. Reckon you might visit us sometime?'

'You bet.' I surprise myself, because deep down I really mean it.

He winks and goes back to the movie.

There's still no sign of the director of nursing when the jobs list has been completed. If I can get out of here without having the pleasure of sharing one last conversation with her, my day will be complete.

I swing open the heavy door to the high care ward for my last official visit with Jack. The fragrant smell of spicy food fills my nostrils and it sounds as if there's a party going on down the hall. Fiona's at her desk looking flustered. This is the first time I've seen her even slightly annoyed. 'Rory, can you tell me what's going on? Who are all these people?'

Down the corridor, groups of people are chatting and coming and going from Jack's room.

'Ah, just give me one minute, okay?'

I push my way through the crowd of people jamming the doorway and I can't believe I didn't see this coming. Everyone from his boxing club is here. It looks like there are even some parents too, all crammed into this tiny space. Out of nowhere vases have appeared, filled with bright flowers. Containers of food are stacked on the bedside table. A helium-filled balloon bobs away in the corner and beside it a huge sheet of cardboard has been stuck to the wall. It's covered with photos of smiling faces. Messages are written in black Nikko pen, some in English but others in languages I don't recognise.

Jack's propped up in bed, grinning from ear to ear. He's wearing a new white t-shirt. Three plastic plates filled with food are laid out in front of him, but it doesn't look as if any of it's been touched. He sees me and holds up a takeaway container as if it's a trophy. The contents look all brown and lumpy, though the smell is amazing.

'Rogan josh!' He's excited.

'And look, you got to keep your right arm, too,' I joke, giving him a big thumbs up.

Part of me is happy for him, but I can't help feeling a bit jealous. This is no longer *our* space. On the other side of the crowded room is Malik. He's watching carefully and a cold shiver passes through me. The room suddenly feels claustrophobic and the smell of the food turns my stomach. I push my way outside and head back to the nurses' station.

And suddenly my day goes from bad to worse. The director of nursing's here and she does not look happy. I could just turn and walk the other way. After all, it's my last afternoon, and the idea of slipping out of here without having to speak to that woman again is definitely appealing. But then I think of Jack and his responsibility speech, and I steer myself straight into the oncoming storm.

'Miss Morris, I might have known you were behind this—' the director waves her arm down the hall in the direction of Jack's room, '—this *situation*.' She takes two steps towards me, her large frame quivering with each move until she's right up in my face. She pauses and sighs, like I'm forcing her to do something unsavoury.

'I've been asking these visitors some questions and it seems you took it upon yourself to take one of my patients out on a field trip, without permission. If you were an employee I would sack you on the spot.'

At this point I can predict what's coming, but instead of flinching, running away, I'm ready to take the hit.

'First thing Monday I will be on the phone to the magistrate. You are aware of what happens if you break the rules of your community service?'

I swallow hard. There was some talk about juvenile detention, of making an example out of me, but as there were no recorded priors, I had filed this information away as *highly unlikely*. But the magistrate did not seem like a man you'd want to cross twice.

'No doubt you will be hearing from him soon. Oh, and your services will not be required here anymore. Either leave now or you will be escorted from the premises.'

She turns, stalks off down the hall and leaves me with only the sound of her shoes clacking on the vinyl floor.

SIXTEEN

I had been on my way home from the shops when I made the detour to Jack's room at St Mary's. I won't let that cow of a director stop me from visiting my friend. Besides, yesterday I didn't get a chance to speak to Jack, and there are things that need to be said.

Fiona's sitting at her desk and the first thing I notice is how pale she looks.

'Rory,' her voice catches as she puts the phone back in its cradle, 'I've been trying to call you.'

That's when I notice her eyes are rimmed with red. An invisible hand slices through the soft flesh of my belly.

Inside Jack's room the morning sun pours through the window. But it feels empty, like he's already gone. But then the sound of his breathing is so loud and I have to resist the urge to cover my ears. Each dragging breath is followed by a gaping silence.

My chair sighs under me as air escapes from the cushion, the vinyl hot against my legs, warmed by the bright sunlight coming through the window. My hands fold together in my lap. A terrible weight bears down, making it hard for me to breathe too. My eyes flick around the corners of the room, looking anywhere but at him. They come to rest on the cobwebs that hang from the edge of the curtain and I watch as they dance in the breeze. The image wobbles and blurs as I scrub angrily at the tears.

On the edge of the bedside table my copy of *Wuthering Heights* lies opened and upside down just where I left it. Its spine is cracked. With trembling hands I pick it up and press the two halves together, cradling it in my lap. The tattered edges of the cover are starting to curl. They feel surprisingly soft, worn by the hands of its readers.

I reach for Jack's hand, to steady myself. It feels cold and weightless already. The skin on his hands is paper thin, covered in liver spots and bruises, and I think about the journey these hands have taken. At one time small and sticky, held and kissed by a loving mother. Hands that have mown yards and made cups of tea. In my mind's eye I can see them over time, twisted by grief and guilt and anger. I squeeze Jack's hand between mine. His knuckles are swollen and bony, crippled by years of throwing punches. Yet these are also the same hands that have been held out, ready to catch lonely souls like Essam and me.

Jack's story is coming to an end.

'Don't leave.' My voice is thin and reedy. It sounds wrong. I've broken a sacred quiet that's descended on the room. Even the ghosts are silent. There's only the terrible scraping sound of his breath.

And then, without warning, it's gone.

My eyes feel scratchy and I rub them with the heel of my hand. Fiona's in the room; her eyes are red, too.

'How long till they come?' I ask.

'The doctor has to write up the death certificate first. Someone from the funeral home is usually here within a couple of hours.' She wipes her eyes again and tucks a tissue into her pocket. 'You'd think I'd be used to this, but it never seems to get any easier.'

There's a quiet knock on the door. A man wearing a long-sleeved shirt and a purple tie breezes in. Fiona greets him by name as he pulls a file from his briefcase. I've seen this guy lurking around. He's in here every day, visiting the oldies, most likely prescribing heart tablets and haemorrhoid cream. I'm sure he would have spent time with Jack but his face is so composed. His eyes hold no sadness. Fiona motions for me to follow her out to her desk. I sit rigid on her chair while she ducks off to make us both a cup of tea. On the desk is a plastic penholder. I reach over, and slowly remove a pencil.

'Fiona,' I say, 'do you have any paper?'

She opens a drawer in her desk and pulls out a few sheets of photocopying paper.

'Will this do?' The doctor is hovering around the nurses' station; he needs to speak to Fiona.

Jack's room feels empty. My chair has been pushed up against the wall; the doctor who came in to write the death certificate had moved it out of the way so he could complete his examination. There was no subtlety about him – all business. To him, Jack was just another old man in a building filled with the aged and the dying.

I lift my chair gently, and return it to its place.

My fingers itch to get started. I wonder briefly if it's possible to forget a part of yourself, but as the first lines move across the page, there's only Jack and me. The pencil whispers its own secret language and, for the first time, I notice things about Jack that, before now, have been invisible to me. The bend in his nose, the faint scar on his left cheek, the patchy grey stubble on his neck, it's as if I'm in a trance, as if no time passes until a hand on my shoulder brings me back. Fiona's standing beside me, and I know my time with Jack is over.

When I step out from St Mary's the sun has well and truly gone down and there is only the glare of street-lights. A few stars struggle to be seen through the artificial light. The walk home has a new urgency. My legs stretch under me, thudding against the pavement. My hands are sweating despite the cool night air and I have to keep shifting my grip on the sketch to wipe my palms. Mum has left the front light on but the rest of the house is quiet. The clock in the kitchen says it's 10.30. Mum's on an early shift tomorrow and she's already in bed. On the kitchen table is a note explaining that Fiona rang to tell her what was going on and that dinner's in the fridge. But the thought of food turns my stomach. A towel sits in a puddle in the corner where it was dropped this morning. It looks as if nothing has changed.

Under the mess in the bottom of my cupboard is my Year 7 art folder. Dad's portrait makes a zipping sound as it's torn free from the sketchpad. I roll it together with Jack's portrait, my hands expertly working the band to secure the two together.

My skin feels itchy with the tension of what's just happened. Outside the streets are quiet: the only sound is the occasional car in the distance. At this hour no one is around. A little voice in my head tries to warn me that walking through the park at night is a really stupid idea, but the pull of my tree is strong. Its roots are deep; it'll hold me steady.

The park's dark except for a light not far from where we sit, casting a strange orange-yellow colour among the leaves on the ground. My tree is close. Shadows move across its rough skin. The tree's root system folds like a giant concertina. My skin prickles. This tree is part of my life, my touchstone, and suddenly I feel silly, almost childish to think that, until tonight, my tree's true purpose has been hidden from me.

I follow the curve of the trunk up to the branches, looking for a way in, through the skin to the heart. My bare feet grip at the giant twisted roots. The tree's rough skin is strangely comforting. My feet and hands follow the folds and curves of one of the high-ridged roots that lead out to the lowest branch. I stop close to the place where our names are carved. Here's the twisted, empty socket I've been searching for. The folded paper is tucked safely inside my t-shirt. Cobwebs fall like a curtain across the small opening only wide enough for my fingers to penetrate. My body is pushed hard against the tree as fingers work to push the folded paper deep enough so that it won't be seen, so that even a child climbing the tree won't know it's there.

I curl my body against the trunk and close my eyes. This is no longer the tree of my childhood.

It's the birds that wake me. Their chorus fills the air with a cacophony of cries. In the grey light of dawn their calls are such an unexpected comfort. I stand and stretch my body in the cool morning air. I reach up and pull off a handful of leaves, crush them in my hands and breathe in their scent. In the grey corner of the sky a few stars continue to flicker. They told us in science that it's technically a planet, but the morning and evening star, Venus, the goddess of love, burns the brightest. The streetlights still glow in the early grey light. What a strange combination this crossroads of day and night is.

I feel dazed, my feet seem to be leading me forward and already I know where they're taking me.

The shed doors are closed. He's not here yet so I sit with my back against the cool of the metal. But it's not long before Essam wheels his bike down the driveway. He raises his hand halfway as a greeting, then stops. My voice breaks over the words and I can't speak through the sobs. He rests his bike against the fence and his bag falls to the ground as he moves towards me. Our arms wrap around each other. He wipes away the long strands of tear-wet hair from my face and holds me till I come to rest.

SEVENTEEN

Sitting inside the church is like being in a giant cave. People huddle in groups of two or three, speaking in hushed voices. It sounds like hundreds of birds' wings as they suddenly take flight. A pair of boxing gloves lies on top of the polished wooden coffin, arranged like hands folded in prayer. The leather is old and stiff, so worn that it has frayed and torn in places. I'm doing my best not to look at them.

A photo of Jack is printed on the front of the order of service. He's standing outside a giant tent, flags fluttering in the breeze behind him, gloves raised, poised in a defensive stance. The thing about funerals is that no matter how well you think you know the person, you'll always discover something new and surprising, and it doesn't take long before I strike gold. In the third paragraph of the program it says that Jack was awarded an Australia Day Achievement Medallion for his work with at-risk youth.

Mum glances over, pats my hand. Fiona, who is sitting in the pew across the aisle, turns around and gives me a smile. I was surprised when Mum said she wanted to come. Maybe it's only because she's working a late shift so she didn't have to ask for time off. Whatever the reason, I'm actually glad she's here.

Essam's sitting not far away with Seraphine, surrounded by other boxing families. Some I recognise, some I don't. Down the back is Rajesh, the cabbie. He sits beside his wife and it looks like his two sons have even made the journey from Brisbane to pay their respects.

The minister comes out and it's as though we all fall under his spell. We stand, we sit, we kneel, we pray. When it's over the pallbearers carry the coffin out of the church with Essam in the lead. It rolls silently into the back of the hearse. The funeral director takes the lead with slow and clipped steps. As he steps out into the street and holds his hand up to stop the traffic, he looks dignified and full of the importance of the moment.

Cars come to a halt and, slowly, the traffic backs up. A young guy wearing a baseball cap is driving a retro-style ute done up with fat mag tyres. It has mounted spotlights and rum stickers across the back window. He rolls down his window and calls out in an angry voice to the funeral director standing in the middle of the street. As the hearse glides into view he works out what's happening and pulls his head in.

Now, driven by some morbid fascination, he turns his gaze on us. Our grief is a sideshow. I want to yell at this yobbo, grab him by the shirt and scream at him to piss off, that he has no right to intrude. What's happening here is important and so much bigger than his infantile brain can understand. Instead, I bite the inside of my cheek.

The hearse turns onto the road and I'm blinded momentarily by a flash of sunlight reflecting off the windows as it disappears past the traffic lights. High up on the church steps the minister stands, watching over us. His robe flutters in the breeze. Essam steps between me and the traffic and envelops me in a hug. For the first time I don't feel at all self-conscious. It feels like the most natural thing in the world.

'We're going back to Jack's shed for the wake. You're very welcome to join us,' he says, and his warm breath tickles my neck.

'Thanks, I might swing by a bit later.'

I head back inside the cathedral to collect my bag. Fiona is talking to the funeral director. I remember back when Dad died that organising a funeral was a bit like throwing a huge party. At very short notice. For everyone – friends and relatives – with the added bonus of it being the crappiest job in the world.

I grab my bag and a couple of extra copies of the order of service.

'Rory, wait up. I've got something for you.' Fiona is striding up the aisle. From inside her coat pocket she pulls out a drawstring bag and places it in my hands. 'Jack made me promise to give this to you.'

I open it and tip the contents out into my hand. Jack's lighter falls cold and heavy onto my palm. It feels solid, real. I read the words again. *Everything happens for a reason.*

Fiona's voice is quiet. 'You were so good for him. It was such a pleasure to see you two together.'

Behind her, light pours through the stained-glass windows. She senses my hesitation and closes my hand around the lighter.

'Oh, he also wanted me to tell you, and I quote, *If she ever smokes, I'll bloody well come back and kill her.*' Fiona's using the low, snappy, gruff voice that is *so* Jack. I sniffle and manage half a smile. We both look down at the lighter and laugh.

EIGHTEEN

The back steps are wet with dew, but I don't mind. My toes are loving the deep lush growth of the lawn. It's still early: too early for Mum and Eddie to be up. My eyes linger on the garage. The lock's starting to rust and I wonder when, if ever, Mum will allow it to be opened again. I try not to think of the spiders that have probably made their home in the pile of stacked canvas. I hope there are no mice. How heartbreaking would it be if mice destroyed Dad's work?

The day's purpose stretches ahead of me. Laid neatly across my bed is my cute blue dress. But this time, the heels can stay in the cupboard. Today is definitely a Docs day. I might look like a fairy with attitude, but at least my feet will be comfy.

At the station, without warning, Mum leans across the front seat of the car and wraps her arms around me in an almost desperate embrace. At first my body is tense, as if I've forgotten what to do, but then something changes and I hug her back. For a brief moment I am eight years old again and safe in my mother's arms.

The sound of the train approaching down the tracks snaps me back to reality. I run down the ramp and onto the platform. The other passengers getting on step forward to open their doors. Mum still sits, watching as I climb aboard and find a seat. Only about half a dozen people are spread throughout the carriage. Through the window I can see that Mum's still in the drop-off zone. We pull away, she holds up her hand, waving, and I watch until the train rounds a bend. Then it's just me and the gentle side-to-side rocking. Stations are called, passengers get on, and slowly the carriage fills. A boy sitting next to me is fiddling with his iPod, scrolling through a list of songs. The tinny bass sound echoes from his headphones.

A couple of stations further down the track two old men, about the same age as Jack, lift expensive-looking racing bikes into the carriage. Fit and wiry, they're clad in lycra that hangs a bit loose in the seat of their pants. Their shirts have a company logo printed across the front. They look like something out of the Tour de France, but for old people. Normally I would scoff and roll my eyes, but they don't look at all foolish. Their wiry frames sit comfortably as they chat to each other, occasionally sharing a joke with one of the other passengers sitting nearby. So at ease with themselves, they laugh and banter like a couple of teenagers. I think of Jack the tent boxer, who always appeared so comfortable in his own skin, and I smile.

My head's pressed against the glass window. For a moment I'm startled at the face reflected back at me, then the train rounds a bend and Mt Tibrogargan looms in the distance. I've always found it hard to believe that these huge eerie shapes were once volcanoes. Now they are nothing more than volcanic cores, extinct for millions of years. Imagine, Dad said once, Captain Cook sailing merrily down the east coast only to look across and see these monsters looming out of the mist. Cook thought they looked like the huge glass furnaces, glasshouses, from his home town of Yorkshire, hence the name. But Dad and I decided to nickname this one the Gorilla. The train snakes towards the Gorilla, so close that I can stare straight into its ancient face. He looks tired. I guess if I'd stood around for millions of years I'd be tired, too.

People continue to come and go until, eventually, a voice announces that we'll be arriving at Central Station. There's a sudden flurry of activity as the commuters put away their laptops and books and the crowd exits, jostling to be the first to be expelled from our metal box. What's the rush?

Looking up to the bank of TV screens that hang from the ceiling I realise my connecting train left five minutes ago and I resign myself to walking. The majority of people milling about in the mall look like worker ants. Dressed in black and grey, heads down, distracted by their phones, showing the world how truly efficient they are. Some of them glance up at me from their takeaway coffees and phones with blank faces. They're wondering what the girl in the Doc Martens and cute dress that swishes so beautifully at her knees is smiling at. And I'm grinning like a Cheshire cat, swinging my bag in a wide arc beside me as I walk.

The giant brown snake that is the Brisbane River rolls and slithers, its scales glistening in the sun. It's so bright that, walking across Victoria Bridge, I'm forced to shield my eyes from the glare. A crew of rowers, pulling and straining, appears, like magic, beneath the bridge. I lean over the railing, watching as their oar blades disappear beneath the brown water and reappear before gliding across the surface. Dip and glide, dip and glide.

Before me stands the art gallery. Its familiar white boxy features trying hard to fit in with twenty-first-century architecture. I run up the steps two at a time and pull at the heavy doors. Inside it's cool, like an oasis, the heat and glare locked out. In the distance I can hear water splashing and I remember the water feature called the Watermall spread across the floor downstairs. There's an information desk in the middle of the foyer. A pleasant-faced man smiles at me. His uniform is perfectly pressed.

'Good morning. Can I help you with anything?'

'No, thanks,' I say confidently. 'I know where I'm going.' But as soon as the words are out there's a pang of doubt. What if she's been moved since I was last here? I double back, clearing my throat, 'Um, can you tell me where *Aurora* is?'

The man wrinkles his sun-bleached forehead, thinking. He's older than I first thought.

'You mean the Burne-Jones painting, *Aurora*?' He points over his shoulder. 'She's exactly where she's always been: just behind us here in the first room.' He leans in as if he's sharing a secret. 'She's one of my favourites, too.'

I'm sure-footed now. She's here, *exactly where she's always been*. How could I ever have doubted it? I round the edge of a display and my skin prickles. I'm face to face with a goddess.

The painting is bigger than I remembered. Aurora towers above me. Cymbals in hand, she glides between stone buildings that rise steeply from both sides of a canal. The early morning sky glows behind the distant rooftops and buildings. *Come forth*, she calls, *come forth into the new day*, and I'm captivated by her movement. She walks barefoot, touching the ground with such lightness. The long, flowing robes gathered above her waist rustle with purpose. The gold light of dawn is reflected in the water of the canal. The stillness of the dawn echoes the sense of solitude in the painting. I know how she feels, walking alone.

Aurora's stare is fixed, her mind elsewhere. Perhaps she's thinking about Tithonus, her mortal lover, who was granted the gift of immortality by Zeus, but not the gift of eternal youth. While Aurora continued to herald the dawn, Tithonus grew old and withered. I remember Dad's contagious enthusiasm as he described details of the artist's love of classical mythology. Edward Coley Burne-Jones; such a strange name for a painter.

Something in the back of my mind is trying to recall an important detail. And then it comes to me. The face reflected back at me in the train. I look more closely at Aurora, this painted goddess, and somehow I know she's with me.

It's Jack's voice that I hear, though: *Believe you me, stumbling across a ghost is not something you forget in a hurry.*

I hardly dare to breathe for fear that this moment will end.

A shrill voice cuts through the air and I'm back, blinking and dazed. There's a tour group gathering behind me and the guide points her clipboard at the painting. 'Does anyone know what role the Roman goddess Aurora plays in classical mythology?' A few hands shoot up as I push my way past them. It's time for me to go. I stop to look back one last time at Aurora.

The escalator glides down to the Watermall. Voices and splashing water echo as if I'm inside a huge underground cave, and I feel an overwhelming urge to get out into the fresh air. A short distance away is the gallery café. Outside, in the courtyard, a massive tree spreads its shade over the tables. A few stray purple flowers hang down among the foliage. The tree's great limbs protect me from the harsh mid-morning light, but I'm still blinking, still wondering what just happened.

A group of older women sits against the courtyard wall chatting over coffee. They're dressed in smart two-piece suits, with photo IDs on lanyards around their necks. One of them has kicked off her heels and is flexing her toes in the sun, enjoying a moment of freedom. Without thinking I search my bag for a pen and start to sketch the women as they laugh and interrupt each other in friendly banter. The serviette fills with bold lines as the pen seems to move on its own, fleshing out the light and shade of these characters. Then a shadow falls across the sketch. The woman who had removed her shoes is standing in front of me. She's studying the sketch, smiling.

'Not bad.' She grins. 'You've done a great job of capturing the moment.'

I feel a sudden surge of confidence and I sign my name down the bottom and hand the serviette to her. She laughs, her eyes shining at my impulsive act. 'Well, Miss,' she pauses to check the signature, '*Morris*, you keep that up and one day you might find one of your works hanging in the gallery.'

NINETEEN

It's Australia Day and I'm busy writing a list of all the things I need to do before school starts in two days' time when my phone goes off. It's a group text from Cam. As I read, my stomach clenches into a knot.

This boatie is going down, flashes across the screen. There's a picture, too. But all I can make out is the back of some kid walking along the road, probably unaware that someone's following them. Another text says that if I want to watch they're at Water Street near the cathedral – now. I can't help it: my fingers touch lightly round my eye. Just breathe, okay, breathe. Panic attack.

A few minutes later there's another message from Cam's phone. *Boat people beware, we're coming for you.* I feel like I want to heave.

Without thinking I shove my feet in my sandshoes and race downstairs, slamming the door behind me. My legs are pumping. There's only the heavy, rhythmic sound of my breath. It's like Jack once said, a sucker punch might win you the fight, but lose you your credibility, the respect of your opponent. And if Cam does this, I am seriously going to lose *all* respect for him.

I stop to catch my breath. From here I can just make out the top of the cathedral spire. There's time for one last nervous glance over my shoulder before passing through the wrought iron gates into the cathedral grounds. The familiar sound of a high pitched laugh echoes and I listen for what seems like forever as it moves off into the distance. I can't wait any longer. As I come around the huge stone building to the back steps there's a sight that stops me in my tracks. There's the body of a boy lying at the bottom of the stairs. Face down. And it's not moving. There's a gash on his forehead and his face is covered in blood. I shake him. Through a haze of panic I'm trying to remember what you do in first aid.

'Hey, can you hear me?'

Head injuries. Was there something about concussion? What about the bleeding? Should I apply pressure? God, I wish I'd paid more attention.

My hands shake as I clumsily prod the side of his neck. I'm holding my breath as I feel for a pulse. I place the back of my hand in front of his mouth and there's a soft, warm tickle: he's breathing. I roll him into the recovery position and he lets out a groan.

I thought I'd be cool in a crisis, that I'd know exactly what to do, but look at me! Then I pull out my phone and dial 000. The pool of blood round the guy's head looks sickeningly red against the sandstone.

The voice on the other end of the line is calm and controlled and somehow I answer the questions without getting impatient. 'Yes, he's unconscious, looks like he's hit his head.' There's a pause. 'Yes, he's breathing and there's a pulse, but I'm not sure how long he's been unconscious. Yes, St Stephen's cathedral on Water Street.'

In the distance sirens wail and the boy opens his eyes. He's probably about 12, jet-black hair. His face is all slack and pale. His lips are thin and white around the edges and his eyes are wide open like he's scared, trying to figure out what's going on. He's starting to shiver and I recognise the first fingers of shock wrapping their tentacles around him.

His voice is groggy and he's speaking in a language I can't understand.

'It's going to be okay, the ambulance is almost here, just lie still.'

When the ambos arrive one of them pats me on the back. 'Okay, mate, we've got it from here.'

Two big men in uniform are bustling around me, reaching for equipment, unzipping bags, asking questions, and I'm ridiculously relieved. Unlike me, these guys know exactly what they're doing and ten minutes later they're wheeling the boy into the back of the ambulance.

One of them turns to me as they shut the back door of the van. 'You did good, kid. Be nice to think there were more good Samaritans like you around.'

I wince inwardly knowing that I'm probably the reason he was hurt.

The lights and sirens are switched on full blast and I stand watching till the ambulance disappears and all I can hear is the siren wailing far off into the distance.

TWENTY

It's all over the local news that night, with a photo of the boy flashed onto the screen. Seeing his face on TV gives me a jolt, making it seem more real. He looks even younger than I remember, probably because it's a cheesy school portrait. The news journalist talks about a spate of racial attacks in town and how two teenage boys have been brought in for questioning. Without thinking, I reach for my phone. But then I stop. And even though there is a sick feeling in my stomach I put it back down again. Mum's sitting next to me on the couch and, when the ads come on, she mutes the TV. She glances over the top of her glasses at me.

'I was there when that boy was brought into Emergency today.'

'And is he going to be okay?'

'He's doing better than expected. You know, I thought that One Punch Can Kill campaign was getting through to people.' She stops to shake her head and I raise my eyebrows, playing dumb. 'It's not the punch that kills; it's the falling unconscious onto a hard surface that does the real damage. He's very lucky; he could easily have been killed, or ended up with a brain injury. Of course, then it's the families who become the primary carers, usually for the rest of their lives, and that's not a burden I'd wish on anyone.' Sounds as if she's on the verge of a full-blown lecture and I think about going to the fridge.

'They say who did it?'

She shrugs. 'The police were in, of course, talking to the boy and asking his family questions.' She puts her coffee mug down and turns to me, and looks me square in the eye. 'Apparently some girl called the ambulance and waited with him.'

'Ha. Hmm.' I shake my head. She's doing that thing she used to do when I was little, trying to stare me down. But I learnt long ago that the best defence against Mum is to just let it pass.

She turns back to the TV, presses the mute button, and the voice of the newsreader fills the room.

We're at our appointment and Mum's beside me in the waiting room as I shift round on the plastic chair, trying to find a comfortable spot. She's had to take time off work to be here, but at least she's here. And even though she has been doing nothing but sigh and stare off into the distance when she thinks I'm not looking, I know she's there for me. The thought is strangely comforting.

The clock on the wall says it's 8.55. Wet patches are spreading under my arms and it's not from the heat. Eddie rummages deep in a Lego box over in the corner. Normally I'd be down there helping him build the tallest tower possible, but right now I feel like I'm about to lose my lunch all over the carpet.

'Miss Morris?' a gruff voice calls down the hall. The magistrate is holding open the door to his office.

We all mooch into the room and Eddie scrambles up onto Mum's lap. There are books and folders everywhere. This time he's wearing a long-sleeve shirt with the sleeves rolled up, and good pants. Without his robes he looks far less scary.

'Aurora Morris, I was hoping that our paths would not cross again, but here we are.' He picks up a stack of paperwork, adjusts his glasses and reads in silence. My heart's hammering in my chest. Is it possible for a 16-year-old girl to have a heart attack? He takes off his glasses, presses his fingers together and leans back in his chair.

'After speaking to the director of nursing from St Mary's last week I must admit to feeling somewhat disappointed. I made my expectations very clear that you were to complete your community service without complication.' He swivels round slightly and looks out the window. 'Miss Morris, you took a resident from the high care ward, *without permission*, I might add.' He swings back, puts on his glasses and reads from the sheet in front of him. 'So you could watch an amateur boxing match at the Nambour PCYC.'

I open my mouth to speak but he holds up a finger.

'Then things get even more interesting. During the week I get these.' He holds up a pile of actual letters. The one sticking out from the pile appears to be handwritten on pink floral paper. 'All of *these* people are trying to tell me what a wonderful person you are. Now, normally I'd write this sort of thing off as a stunt. But I have here letters from a staff member *and* a resident of St Mary's, members of the Sunshine Coast Amateur Boxing Club and last, but certainly not least, a letter from the family who own the Indian restaurant you and your hooligan friends trashed.' He's speaking as if he truly can't believe what he's just said. He spreads the letters across the table in front of him and Mum grabs at my arm. She's pressing it so hard I have to bite the inside of my cheek not to yell at her.

'I'm not sure what's brought about such a huge change in your attitude, but I do feel that in the face of such overwhelming evidence, it's safe for me to say that I will never, *ever* see your face inside my court again. Would I be right in that assumption, Miss Morris?'

I'm totally speechless. All I can do is nod.

As we file out of the office he puts his hand on my

shoulder. 'Miss Morris, could you also tell your friends that as much as I love a good curry, they can stop sending me food. I am not permitted to accept gifts of any kind, regardless of how good they smell.' He turns to go but glances back one last time. 'Oh, and Rory?'

'Y – yes.'

'Have a great life.'

I wake early the next morning thinking about Essam and Jack and how much has changed since I met them. And the thought of meeting up with Essam later today brings a smile to my face.

Eddie's still asleep in his bed, his arm wrapped tightly around his bear. I pull the sheet up to his chin and he murmurs softly and rolls over, tangling himself in the sheets. Mum's room is empty, her bed already perfectly made. Tight hospital corners have always been her specialty, but her queen-sized bed looks too big for one person.

There's a noise outside and, when I pull the curtain across, my heart jumps. The orange glow of morning shines through the window. Is it lack of sleep, or am I seeing things? The garage door's open. A figure steps out onto the lawn. Mum, dressed in her cut-off overalls, her cleaning clothes, dust cloth in hand. She shakes the cloth and a puff of dust escapes. Her hand covers her mouth and she turns her head away. For a moment I feel anger. How dare she intrude on Dad's space? But after a moment it passes. She looks up at me standing at the window. On her face is a mix of surprise and fear, like she's seen a ghost, and she stumbles out of the shed onto the lawn. As she turns back towards the window her feet leave a mishmash of dark tracks on the grass. She raises her arm to shield her face from the sun. There's a tentative wave and smile.

I wave back, as if to a friend going past on the street. There's a strange excitement. A world of possibilities opens. Yes, death changes the living; why wouldn't it? But right now it's time to make the coffee. And it's me who boils the kettle, stirs in the milk and sugar, and wipes the bottom of the carton where a drop has dribbled down. I head out to the garage, balancing the two mugs as steam swirls around them. The early morning dew is cool against my feet and no doubt the birds will be back sometime soon.

ACKNOWLEDGEMENTS

Firstly, I would like to thank Nike Sulway who was there from the beginning and who taught me how to write a scene. Without your kind words, wisdom and enthusiasm, I would never have had the confidence to attempt something as impossible as a novel.

Secondly, thank you to Kristina Schulz for your excellent editorial advice and warm welcome to UQP, and to the lovely Kristy Bushnell, who put the 'Becoming' in *Becoming Aurora* – I feel so fortunate to be working alongside both of you and the entire team at UQP.

Next, thank you to Mark Macleod, whose keen eye and attention to detail helped transform my sentences into something shiny and beautiful and to Jo Hunt for the gorgeous cover design (it is truly, truly awesome). And to Christine Bongers, thank you for your wonderful endorsement of the book.

Thanks also to the judges of the Queensland Literary Awards. I will always be grateful to you for taking a chance on my manuscript.

I'm grateful to Varuna House for the gift of two weeks thinking and writing time, and to the awesome women writers I met there – Jane Rawson, Amanda Webster, Melissa Beit, Arianne James and (in particular) Demet Divaroren – who showed such interest and enthusiasm in this story.

To my first readers – Jenny Carrington, Lorena Carrington, Alison Condliffe and Anne Tyndall – thank you for your enthusiasm, and to Alison Quigley and Taryn Bashford, thank you for your insightful and timely critiques.

To St Luke's, thank you for allowing me the use of the room in the back of the church so I could edit away from the distractions of the internet.

And lastly, to my family, and in particular to my mum, Kath Haworth, thank you for everything and for always being there. And to James (I know you don't want to be thanked but I'm doing it anyway!), you held the fort while I hid away from the chaos to write. And to my boys, Will, Rhys and Alec, who would track me down to ask: *What's to eat?*, I love you munchkins more than you know.

ONE WOULD THINK THE DEEP
Claire Zorn

From the multi award-winning author of *The Protected* and *The Sky So Heavy* comes a ground-breaking young adult masterpiece about lost young men.

Sam stared at the picture of the boy about to be tipped off the edge of the world: the crushing weight of water about to pummel him. He knew that moment exactly, the disbelief that what was about to happen could even be possible. The intake of breath before the flood.

Sam has always had things going on in his head that no one else understands, even his mum. And now she's dead, it's worse than ever.

With nothing but his skateboard and a few belongings in a garbage bag, Sam goes to live with the strangers his mum cut ties with seven years ago: Aunty Lorraine and his cousins Shane and Minty.

Despite the suspicion and hostility emanating from their fibro shack, Sam reverts to his childhood habit of following Minty around and is soon surfing with Minty to cut through the static fuzz in his head. But as the days slowly meld into one another, and ghosts from the past reappear, Sam has to make the ultimate decision … will he sink or will he swim.

'A beautifully crafted novel about love lost and regained, about families and the secrets that tear them apart, and which finally hold them together. Claire's previous books, were award winners. With *One Would Think the Deep*, she has raised the bar even higher.'
Bill Condon, award-winning author of
A Straight Line to My Heart

ISBN 978 0 7022 5394 2

ANOTHER NIGHT IN MULLET TOWN
Steven Herrick

An iridescent verse novel from award-winning author Steven Herrick that shines light on friendship, family and finding your way.

'People like you and me, Jonah,
we drag down the price of everything we touch.'

Life for Jonah and Manx means fishing for mullet at the lake, watching their school mates party on Friday night and wishing they had the courage to talk to Ella and Rachel.

But now their lakeside town is being sold off, life doesn't seem so simple. Manx holds a grudge against the wealthy blow-ins from the city and Jonah just wants his parents to stop arguing.

One memorable night at the lake will change everything.

'Herrick's poetic approach really gets to the guts of the story.' *Books+Publishing*

'Herrick is an expert writer.' *Weekend Australian*

ISBN 978 0 7022 5395 9

FRANKIE AND JOELY
Nova Weetman

'Joely,' says Frankie. 'Before it starts ... this holiday ... it's about us. Right?'

Frankie and Joely are best friends. They love each other like no one else can.

It's summer and, together, the girls are escaping the city and their mums for a week of freedom in the country. But when Joely introduces Frankie to her country cousins, Thommo and Mack, it soon becomes clear that something other than the heat is getting under their skin.

As the temperature rises, local boy Rory stirs things up even more and secrets start to blister. Will they still be 'Frankie and Joely' by the end of their holiday?

'A quietly mesmerising, finely wrought tale of friendship, loyalty, loneliness and "firsts". *Frankie and Joely* went straight to my heart.' Simmone Howell, author of *Girl Defective*

'I don't think I've read another young adult novel in which female friendship is dissected with such skill and with such attention to detail.' *Australian Book Review*

ISBN 978 0 7022 5363 8

TALK UNDER WATER
Kathryn Lomer

Isn't anyone who they say they are anymore?

Will Lane never expected to make friends with some random girl on the internet.

Summer Rainbird never expected some random guy from the internet to turn up on her doorstep.

Both Will and Summer are missing a parent and needing a friend. But when they meet in real life, will they even be able to talk to each other, let alone be friends?

'This book was genuinely difficult to put down … a reading experience that leaves you with a warm glow.' *Magpies*

'A sweet and charming novel about friendship, trust and finding your way in the world. [Lomer's] exploration of the deaf community adds a fascinating dimension to the story.' *Books+Publishing*

ISBN 978 0 7022 5369 0